Visualizing Chemistry

Visualizing Chemistry

Investigations for Teachers

Julie B. Ealy and James L. Ealy, Jr.

The American Chemical Society, Washington, DC 1995

Library of Congress Cataloging-in-Publication Data

Ealy, Julie B.
    Visualizing chemistry: investigations for teachers / Julie B. Ealy and
James L. Ealy, Jr.

        p.   cm.
    Includes indexes.

    ISBN 0-8412-2919-8

    1. Chemistry—Experiments.

    I. Ealy, James L. (James Lee), 1943-   . II. Title.

QD43.E25 1995
542—dc20

95-43109
CIP

We dedicate this book to our parents,
Forest (deceased) and Evelyn Bowen
and James and Catherine Ealy, for a lifetime
of encouragement and to our children,
Georgia Anne Fisher and James Ealy, III.

1995 Advisory Board

Contents

Organic and Biological

Appendices

Subject Index

About the Authors

JULIE B. EALY received her B.A. from the State University of New York at Buffalo and her M.S. from Northern Illinois University. She has taught science courses from earth science to physics, including honors and advanced placement (AP) chemistry, for 24 years in public and private schools. She is now pursuing a Ph.D. at Teacher College, Columbia University. She develops self-contained demonstration kits for teachers, marketed by Flinn Scientific Company. She writes for the ACT test and has found a safer substance, glyoxal, to substitute for formaldehyde in several popular classroom demonstrations.

JAMES L. EALY, JR., received his B.A. in chemistry and biology from Shippensburg University and his M.Ed. from Lehigh University. Since that time, he has taught at Mercersburg Academy and The Hill School, and now teaches at The Peddie School in Hightstown, New Jersey. He serves as a consultant in AP chemistry for the College Board and has served as a member of the ACS Advanced Test Committee II. Jim is a 1993 Presidential Award winner in science and math. He received a Toyota Tapestry grant in 1995 to use CBLs in radio-controlled model gliders. He is an avid radio-control modeler and was previously the editor of the Vintage Sailplane Association's journal, *Bungee Cord.*

Together, the Ealys have presented numerous workshops on chemical demonstrations to elementary school through college teachers at many local, national, and international conferences. They previously designed the chemistry programs for the Johns Hopkins Center for Gifted and Talented Youth. Both have served as AP chemistry readers and were members of the American Rewrite Committee for the very popular Canadian text, *ALCHEM*. During the 1989–1990 academic year, Jim was a visiting fellow at Princeton University doing educational research on laboratory effectiveness supported by a grant from the National Science Foundation (NSF), while Julie was one of four high school teachers who carried out original research in their own high schools supported by the same NSF grant. The Ealys had the opportunity for two summers to do chemical research on host–guest compounds at Franklin and Marshall College and have papers published from their research. They both evaluate and write questions for the National Teacher Exam published by the Educational Testing Service. They coauthored the manuscript and were demonstrators for the video *Close-Up on Chemistry*, produced by the American Chemical Society. Jim and Julie are coauthors of *Chemical Demonstrations: A Sourcebook for Teachers*, Volumes I and II, respectively, also published by the American Chemical Society.

Preface

Maybe once in your life an opportunity presents itself. This undertaking was our opportunity, and fortunately we were able to recognize it. The exact moment when the germ of an idea to write a book about demonstrations has been lost, and maybe for good reason. Any project of this magnitude requires that the authors be best friends. It is said, "You should never room together with your best friend, because the very best that can happen is you remain best friends." We learned from previous collaborations that we complement each other in many, many ways. Just as children carry the genes of both parents, this book is the result of a wonderful and fulfilling scholarly union. All the trials and tribulations associated with child rearing can be a metaphor for the writing of this book.

We are indebted to our parents for starting this whole process. Evelyn Bowen, a nurse, took her daughter's stitches out in the kitchen and taught her how to light a gas stove. This light is still carried and passed on to her students. Forest Bowen took his daughter hunting and showed her how to clean the animals, and got her up in the middle of the night to watch a lunar eclipse. He forever instilled a curiosity that has left her childlike. Jim, Sr., and Catherine Ealy provided a farm with each day a new laboratory experience. They also provided a chemistry set for Christmas, which Jim, Jr., soon outgrew and began synthesizing compounds not supplied. Above all else, our parents provided safe havens in which we could experiment, take risks, and learn about nature and science. Also, both of us had one or more teachers who made a difference in ways that have yet to be explained in a coherent and satisfactory manner by educational theorists.

Foremost in our minds was an intense desire for this book to be scholarly. We soon discovered how difficult it was to do adequate research without an excellent library and access to Chemical Abstracts Service. In this respect we are indebted to the staff of the chemistry library in Frick Labs, Princeton University. An undertaking of this type is not usually done by high school teachers, and it did not take us very long to discover why. "Hobbits" in the form of graduate students were not available to try new procedures, prepare solutions, and do other forms of grudge work. During one particularly intense period, our daughter and several of her friends asked, "When you go out for a candlelight dinner, do you talk about chemistry?" During the writing of this book we talked chemistry in more places than at the table—some of those places and times might be considered unusual.

We are indebted to Claude Yoder of Franklin and Marshall College for providing advice and help with several demonstrations. Especially when Julie broke a melting point apparatus, Claude handed Julie a screwdriver and said, "Why not fix it?" This was a very respectful expectation and proved to be a turning point in her career. Jim Spencer of Franklin and Marshall provided the framework for the maturing of a teaching style in Jim that we hope is apparent in the investigations that he wrote. Also, Michael Bentley at University of Maine at Orono provided Julie with lab space and chemicals and inculcated more self-confidence in her. We are also indebted to Ray Oram, Science Department Chair, The Peddie School, who provided material help in the form of a room and funds. More importantly, he provided an intellectual climate for our new ideas and discoveries. We are also thankful for the assistance of two of our students. Both students "teacher proofed" almost every demonstration. We, in return, provided them an opportunity to do independent research during their senior year at Peddie. Jennifer Miksis is now a science major at Harvard, and Nicole Gerson is a science major at Franklin and Marshall College.

We are indebted to Cheryl Shanks, American Chemical Society Books Department, for support and trust. She is without doubt intellectually critical, considerate, and the generous editor a project of this scope needed. When we needed more time, she was there with considerable patience and support. The reviewer who checked each demonstration was a blessing to us in every way. To have a critical reviewer who can take your most favorite and cherished "child" and critically make it over requires incredible tact, wit, and an intellectual sense of humor. To that person, we are grateful and respectful, as it is no easy task to criticize another's offspring. Much improvement is the result of his or her beneficent manner, generous time, and thorough commitment to this project. Any and all mistakes, confusion, or lack of clarity is solely our responsibility. We feel the format is helpful and easy to use, and we hope it will make each demonstration inviting and easy to use safely. We wish to express our heartfelt thanks to Keith Ivey for his editorial and chemical knowledge in helping to complete the final review of the manuscript. We are exceedingly grateful that the American Chemical Society saw fit to trust a major project to two high school teachers, and we hope we have fulfilled that trust.

We have since childhood taken "the road less traveled", not often easy and less often recommended, but it is by far the most interesting and exciting. We hope we have cleared some of the brush, and left a clean and unspoiled landscape; in that respect we wish to share our journey with you. Sharing is the highest and most generous compliment we can give you, the reader.

Physical Properties

Hardness and Lattice Energy

The hardness of a mineral refers to its relative resistance to scratching. Although hardness is not precisely defined quantitatively, a relationship exists between hardness and lattice energy, which is the energy released when oppositely charged ions come together to form a solid.

Procedure

1. Display a collection of minerals with their names.
2. Provide students with data concerning the hardness and lattice energy of the minerals.

Concepts

▲ Minerals are often identified by properties such as hardness, color, external shape, luster, cleav-

Materials

▲ Mineral collection: aluminum oxide (Al_2O_3), barium carbonate ($BaCO_3$), cadmium sulfide (CdS), copper(I) sulfide (Cu_2S), lead(II) sulfide (PbS), magnesium carbonate ($MgCO_3$), manganese(II) sulfide (MnS_2), potassium chloride (KCl), silver sulfide (Ag_2S), sodium chloride (NaCl), strontium carbonate ($SrCO_3$), titanium oxide (TiO_2).

age, and specific gravity. A mineral can also be chemically analyzed to determine its crystal structure.

▲ Any solid body that grows with planar surfaces is a crystal. The word "crystal" originated from what the Greeks called ice, *krystallos*; the Romans latinized the name to *crystallum*.

▲ In the 17th century the crystal faces, or planar surfaces that bound a crystal, were found to differ widely in size from one sample to another for the same mineral. In 1669, Nicolaus Steno, a Danish physician, demonstrated that a particular mineral is unique because of the angles between the faces, not the relative face sizes.

▲ In 1912, the German scientist Max von Laue used X-rays to demonstrate that atoms are packed in fixed geometric arrays in crystals. When mineral grains grow freely in an open space, crystals can form.

▲ Hardness is defined as the resistance a smooth surface of a mineral offers to scratching. It is a distinctive property of minerals and is dependent on the crystal structure and the strength of the bonds between atoms.

▲ The stronger the binding forces between its atoms, the harder the mineral. The strength of forces holding the atoms together differs in different directions. Crystals may show varying degrees of hardness depending on the direction in which they are scratched. This fact puts some limitations on determining the hardness.

▲ *Mohs's relative hardness scale,* which is divided into 10 steps, each marked by a common mineral, is used to determine hardness. The softest mineral known is talc (hardness 1 on Mohs's scale) and the hardest is diamond (hardness 10). Any mineral on the scale will scratch all minerals below it, and minerals on the same step are just capable of scratching each other. A pocketknife or a piece of plate glass has a hardness of 6, a copper penny has a hardness of 4, and fingernails have a hardness of 3.

▲ Lattice energy is the energy released when oppositely charged ions in the gaseous phase come together to form a solid. For example,

$$Na^+(g) + Cl^-(g) \rightarrow NaCl(s) + 787 \text{ kJ}$$

▲ Lattice energy depends on the charge of the ions and distance between centers of ions when they pack to form a crystal.

▲ If the value for lattice energy is not known, it can be estimated by using the Kapustinskii equation:

$$E = 108{,}000 Z_c Z_a v/r$$

where E is lattice energy; Z_c and Z_a are absolute integral values of charges on the cation and anion, respectively; v is the number of ions in the simplest formula, and r is the sum of the ionic radii. When r is in picometers, E will be in kilojoules per mole.

▲ The formulas, common names, hardnesses, and lattice energies for three groups of minerals that show a relationship between hardness and lattice energy are as follows:

Formula	Common Name	Hardness	Lattice Energy (kJ/mol)
KCl	sylvite	2	698
NaCl	halite	2.5	769
Cu_2S	chalcocite	2.5–3	2483
CaF_2	fluorite	4	2611
TiO_2	rutile	6–6.5	9031
Al_2O_3	corundum	9	15,326
$BaCO_3$	witherite	3.5	16,000
$SrCO_3$	strontianite	3.5–4	18,305
$MgCO_3$	magnesite	3.5–5	30,000
Ag_2S	argentite	2–2.5	2167
PbS	galena	2.5	2851
CdS	greenockite	3–3.5	3097
MnS_2	alabandite	3.5–4	3402

Notes

1. If earth science or geology is taught in your school, you could have students test the hardness of the minerals.

2. Provide students with the ionic radii of some of the ions and have them calculate the lattice energy using the Kapustinskii equation.

3. See the References for several geology texts that might be useful (Keller, 1982; Larson and Birkeland, 1982; Skinner and Porter, 1992).

4. A useful extension of this demonstration would be to use an overhead projector, placing on it transparent materials such as Formica, plate glass, Pyrex glass, and plastic and determining the hardness of the substances. Quartz, with a hardness of 7, could be used to test the substances as well as a pocketknife, copper penny, and fingernail.

References

Bodner, G. M.; Pardue, H. L. *Chemistry, An Experimental Science;* Wiley: New York, 1989; pp 262–263.

Dana, J. D. *Dana's Manual of Mineralogy;* Wiley: New York, 1963; pp 152–153.

Huheey, J. E. *Inorganic Chemistry*, 3rd ed.; Harper & Row: New York, 1983; pp 59–71 (lattice energy values).

Keller, E. A. *Environmental Geology*, 3rd ed.; Charles E. Merrill: Columbus, OH, 1982.

Larson, E. E.; Birkeland, P. W. *Putnam's Geology*, 4th ed.; Oxford University: New York, 1982.

Shannon, R. D. *Acta Crystallogr. Sect. A* **1976**, *32*, 751 (ionic radius values).

Skinner, B. J.; Porter, S. C. *The Dynamic Earth;* Wiley: New York, 1992.

The Floating Egg

An egg placed in a saturated sugar solution and then in water floats in one and sinks in the other.

Procedure

Please consult the Safety Information before proceeding.

1. Place 400 mL of water in one beaker and 400 mL of the saturated table sugar solution in a second beaker.

2. Place an egg in each beaker. An egg sinks in the water and floats in the sugar solution.

Safety Information

Although food substances are used, no substance should ever be tasted in the laboratory.

Concepts

▲ Because an egg is denser than water, it sinks.

Materials

▲ Sugar solution, $C_{12}H_{22}O_{11}$: Warm 400 mL of water on a hot plate–stirrer and dissolve 145 g of table sugar.

▲ Two eggs.

▲ When a saturated solution of sugar is prepared, the density of the solution is greater than that of an egg, and therefore the egg floats.

Notes

1. Corn syrup, which is also dense enough, could be used in place of the saturated table sugar solution.

2. Sodium chloride has a low solubility in water, and its solubility does not increase very much even when water is heated. A saturated solution of salt is difficult to prepare; therefore, sugar is used instead.

3. Eggs that are no longer fresh will, in fact, float, and farmers sometimes check their eggs this way. Leaving an egg unrefrigerated for approximately 4 weeks results in an egg that will float in water. The egg will have lost about 3 g in mass at this point. It is a good idea to keep track of the weight loss daily. Keeping a record of this and sharing it with the students allows you to extend the discussion of density. Also, make sure to put the egg in water daily to check its progress.

References

Lippy, J. D., Jr.; Palder, E. L. *Modern Chemical Magic;* Stackpole Co.: Harrisburg, PA, 1959; p 104.

Absorption of Laser Light

When two laser beams of differing wavelengths are directed through a light blue solution, one of the beams does not pass through.

Procedure

Please consult the Safety Information before proceeding. Also consult laser safety precautions that come with the instruments.

1. Fill two test tubes one-third full with Vanish solution.

2. Place the tubes in two different racks, one behind the other, so that you can direct the beams of the two lasers through both tubes onto a piece of white poster board (see Notes).

3. Notice that the diode laser beam passes through both tubes and the helium–neon laser beam does not. Place an index card between the two

Materials

▲   Helium–neon laser of 632.8 nm and a diode laser of 670 nm and both 0.8 ± 0.2 mW of power.

▲   Sulfuric acid, concentrated stock solution.

▲   Aqueous ammonia, 6.0 M NH_3(aq): Add 40.5 mL of concentrated aqueous ammonia to enough water to make 100 mL of solution.

▲   Vanish: Add 5.0 mL of Vanish liquid to 1 L of distilled water. Vanish is a cleaning solution sold in supermarkets.

▲   Erioglaucine, C.I. Acid Blue 9, FD&C Blue No. 1: Add 20 drops of blue food coloring containing FD&C No. 1 to 250 mL of distilled water.

tubes to show that the He–Ne beam does not pass through even the first test tube.

4.   With stirring, slowly add drops of concentrated sulfuric acid to the first tube until the solution turns green and then yellow. Place the index card between the two tubes to show that both beams pass through the first tube. Remove the card.

5.   With stirring, slowly add drops of acid to the second tube until it also turns yellow. Both beams should now be shining onto the white poster board. Slowly add drops of aqueous ammonia to the second tube until the He–Ne laser beam disappears and the color returns to blue.

6.   Repeat Steps 1–6, with the erioglaucine indicator solution in place of Vanish.

Concepts

▲   The wavelength of the diode laser light is 670 nm, and this wavelength is not absorbed by the blue form of the indicator. Thus, the laser beam passes through.

▲   The wavelength of the He–Ne laser beam is 632.8 nm, and this wavelength is absorbed by the blue form of the indicator. Thus, the beam does not pass through the test tube.

▲   When acid is added to the blue solution, the pH decreases, the indicator changes to the yellow form, and the solution no longer absorbs in the 632.8-nm range.

▲   When base is added, the pH rises and the color of the indicator returns to blue. The solution now absorbs in the 632.8-nm range. The small difference in wavelength is enough to be selected for or against by the different structural forms of the indicator.

▲   From the structures under Reaction, the H^+ ion present in excess in the acid solution would likely attach to the negative site of the $-SO_3^-$. This bonding would change the character of the structure and change its absorbance. The intermediate green color results from the combined effect of the yellow form and the blue form, both present in the test tube.

▲ This demonstration is an example of color produced as a result of light absorption by an organic molecule. The more familiar phenolphthalein (refer to Investigation 75) in the leuco form (pH < 9.3) absorbs in the ultraviolet region and appears colorless. However, when this *triphenylmethane* dye is subjected to a pH above 9.3, the structure becomes fully conjugated and now absorbs in the 550-nm range (green) and the solution transmits red. Much of the light energy absorbed is converted to heat energy. Simple modification with an oxygen bridge to the phenolphthalein molecule gives the fluorescein molecule (refer to Investigation 90), which absorbs in the ultraviolet range and reradiates most of this energy in the visible range as a deep green *fluorescence*.

Reactions

Vanish, erioglaucine (blue)   erioglaucine (yellow)

Notes

1. Mount the diode laser on top of the He–Ne laser with double-sided foam tape so both are shining in the same direction and through the two tubes. Use books or small bean bags to support the two lasers. Use large test tubes in an open test tube rack for ease of operation. Place a piece of heavy poster board next to the second rack and tilted at an angle, so the students can see the beams from their seats. You may have to darken the room.

2. Be sure to do Step 6 to demonstrate that the blue coloring—erioglaucine—is in Vanish.

3. You may have to adjust the concentration of the Vanish for your laser beam to be completely absorbed in the particular size test tube that you use. If you use a laser of a differing wattage than we used, you may also have to adjust the concentration.

References

Brecher, K. *The Physics Teacher,* Oct. 1991, pp 454–456.

Goelz, J. The Drackett Products Company, Chemistry Department, personal communication, 1992.

Nassau, K. *The Physics and Chemistry of Color;* Wiley-Interscience: New York, 1983; pp 125–127.

Determination of Molecular Weights

The molecular weights of different gases can be determined by using a tube of known length and a Stanley Estimator. Sound waves travel at different speeds in gases of different molecular weights. By using a Stanley Estimator, the apparent "length" of the tube can be determined. A different length is indicative of the unique speed of sound in each gas and is proportional to the gas's molecular weight.

Procedures

Please consult the Safety Information before proceeding.

1. Seal one end of a 2-m-long tube. Place the Stanley Estimator at the open end and obtain the distance for air.

2. Place a different gas in the tube by filling it from a commercial gas tank and repeat the measurement.

Safety Information

1. Carbon dioxide: Dry ice can cause frostbite and death if the gas is inhaled. Handle the dry ice with thermal gloves and work in a well-ventilated area.

2. Oxygen gas is an excellent oxidant, and extreme care should be exercised when using it. Use oxygen gas in a well-ventilated area with no open flames or combustibles.

3. Helium does not present a danger in a well-ventilated room. Do not allow anyone to breathe in helium. (Youngsters enjoy the effect on their voices, but it is not advisable to allow this.) →

4. Both hydrogen and methane gas should be used with extreme caution. Both are highly flammable and should be used only in a well-ventilated room. Do not use open flames.

3. Repeat with as many gases as you have available and record the length of the tube.

4. Substitute the new "length" into the ratio of $MW_{air}/MW_{gas} = Length_{air}/Length_{gas}$ and solve for the molecular weight of the gas. Compare the experimental and actual values.

Concepts

▲ The speed of sound in dry air varies with temperature, but it will be about 340 m/s at classroom conditions. The Stanley Estimator uses this phenomenon to determine the time for the emitted sound wave to strike the object (in our case, the end of the tube) and return to the sensor. The circuitry in the Stanley Estimator converts elapsed time to a numerical number representing distance in feet.

▲ When the air is replaced with a different gas, the sound takes more or less time to travel the fixed distance, and the timing chip interprets this as a different distance. Thus, the velocity can be deduced from this information.

▲ The molecular mass of a gas and its subsequent average molecular velocity are inversely related. Therefore, the speed at which the sound wave can travel through a gas is inversely related to the average molecular weight of that gas. When you use the speed–distance direct relationship, the molecular weights of different gases can be obtained with good agreement. A typical example is as follows:

Gas (dry)	V_{act} (m/s)[a]	$Length_{expt}$ (ft)	MW_{expt} (g/mol)
Air	340	6.6	29
Carbon dioxide	259	10.2	45
Methane	430	3.7	16
Nitrogen	334	6.6	29
Oxygen	316	7.3	32
Hydrogen	1284	0.6	2.7

[a]V_{act} is actual speed of sound at 25 °C.

Notes

1. A good source for a tube is one that wall charts come in or tubes from a carpet shop. The tube can be as small in diameter as 1.5 in., but a tube 4 in. in diameter works best. Poly(vinyl chloride) (PVC) plumbing pipe is excellent and will cost about $5.

2. The major problem is getting the new gas as pure as possible in the tube. If you start with carbon dioxide, immediate success is more likely. Place about 300 g of dry ice in the tube and cover the open end with cloth. The tube should be placed in an upright position.

3. After the dry ice has sublimed, and because it is heavier than air, it will gradually replace the lighter air. If care is used, the tube will be filled with almost pure carbon dioxide gas. If you do not wait long enough, the gas in the tube will be too far below room temperature to give accurate readings. When the bottom of the tube is no longer cold to the touch, take your readings.

4. Other gases include hydrogen, nitrogen, oxygen, helium, and argon available from gas supply houses. For all gases except hydrogen and helium, place the tube upright with the closed end down. Place the gas delivery tube in the open end, down into the tube as far as possible. Gas is slowly released into the tube to gradually replace the air. Oxygen (32 g/mol vs. about 29 g/mol for air), argon (40 g/mol), and nitrogen (28 g/mol), all nearly as heavy as or heavier than air, are also best done the same way. Helium and hydrogen should be done by placing the closed end of the tube at the top and filling the tube by displacing the air downward. Hydrogen is very good because the speed of sound in hydrogen is about 4 times that in air, and the tube appears to suddenly have shrunk. The molecular weight of hydrogen will be more accurate if you use a longer tube, about 10–12 ft long.

Materials

▲ Stanley Estimator.

▲ Commercial tanks of various gases: carbon dixode (CO_2), methane (CH_4), nitrogen (N_2), oxygen (O_2), hydrogen (H_2).

▲ 2-m tube, cardboard or PVC plumbing pipe (see Notes).

References

CRC Handbook of Chemistry and Physics, 66th ed.; Weast, R. C., Ed.; CRC: Boca Raton, FL, 1985.

Reactions of Some Elements

Copper Crystals on Phosphorus

When a piece of white phosphorus is placed in a blue solution of copper(II) sulfate, brown crystals of copper begin growing on the phosphorus.

Procedure

Please consult the Safety Information before proceeding. Handle white phosphorus only in a fume hood.

1. Select a test tube that can be completely filled with copper(II) sulfate solution and a stopper to fit.

2. Fill the test tube with 1.0 M copper(II) sulfate solution, leaving only enough air space for the stopper.

3. Place a cleaned, pea-sized piece of phosphorus in the test tube. Stopper the test tube.

Materials

▲   Copper(II) sulfate penta-hydrate, $CuSO_4 \cdot 5H_2O$, 1.0 M: Dissolve 25 g in enough water to make 100 mL of solution. 0.1 M: Dissolve 2.5 g in enough water to make 100 mL of solution.

▲   Under water, with adequate ventilation, wearing gloves, and handling with forceps, scrape the outside coating off two pea-sized pieces of white phosphorus. Leave the pieces under water until ready to use.

▲   Ammonium molybdate, $(NH_4)_2MoO_4$: Carefully add 10 mL of concentrated nitric acid, HNO_3, to 3 mL of water. Dissolve 0.2 g of $(NH_4)_2MoO_4$ in this solution.

▲   Nitric acid, concentrated: commercial solution.

4.   Observe the copper crystals that will form easily within a class period.

5.   Repeat with 0.1 M copper(II) sulfate solution.

6.   Leave both test tubes out where observations can be made over a period of 1 week.

7.   After 1 day, remove one pipetful of solution from each test tube and place them in separate test tubes.

8.   Place 1 drop of concentrated nitric acid and one pipetful of filtered ammonium molybdate in each test tube. Swirl to mix the contents. Within about 10 min a yellow tint will develop at the bottom of the test tube and a yellow precipitate in about 24 h.

Concepts

▲   In the reaction, phosphorus is oxidized to P^{5+} by losing electrons. Cu^{2+} gains electrons and is reduced to Cu metal. The half-reactions and reduction potentials are as follows:

$$10Cu^{2+}(aq) + 20e^- \rightarrow 10Cu^0(s) \qquad 0.337 \text{ V}$$

$$P_4(s) + 16H_2O(l) \rightarrow$$
$$4H_3PO_4(aq) + 20H^+(aq) + 20e^- \quad 0.411 \text{ V}$$

The positive electromotive force value, 0.748 V, indicates that the reaction is spontaneous at room temperature and requires no additional input of energy.

▲   Because phosphorus is oxidized, it acts as the reducing agent. The oxidizing agent is Cu^{2+}, which is reduced.

▲   Ammonium molybdate is used in qualitative analysis to test for the presence of PO_4^{3-}.

Reactions

1. $P_4(s) + 10Cu^{2+}(aq) + 16H_2O(l) \rightarrow$
$$4H_3PO_4(aq) + 10Cu(s) + 20H^+(aq)$$

2. Production of ammonium phosphomolybdate precipitate:

$$PO_4^{3-}(aq) + 12MoO_4^{2-}(aq) + 3NH_4^+(aq) + 24H^+(aq)$$
$$\rightarrow (NH_4)_3PO_4 \cdot 12MoO_3(s) + 12H_2O(l)$$
$$\text{yellow}$$

Notes

1. The 1.0 M solution will retain a blue tinge, whereas the 0.1 M solution will become colorless.

2. The crystals that grow on phosphorus in 1.0 M $CuSO_4 \cdot 5H_2O$ are shiny and quite beautiful. The crystals in 0.1 M $CuSO_4 \cdot 5H_2O$ are very dull.

3. A good time to set up the test tubes might be before discussing oxidation–reduction.

4. A bottle filled with copper(II) sulfate solution could also be used. Use a larger, more observable piece of phosphorus.

5. A 40-mL test tube works well for the demonstration.

References

Bryce, W. A. *J. Chem. Educ.* **1958**, *35*, A267.

Greenwood, N. N.; Earnshaw, A. *Chemistry of the Elements*, 1st ed.; Pergamon: New York, 1984; p 590.

Masterton, W. L.; Slowinski, E. J.; Stanitski, C. L. *Chemical Principles*, alternate ed.; Saunders: Philadelphia, PA, 1983; pp 736–740.

Charcoal Decolorizes

Black charcoal is added to a purple solution of methyl violet. Upon filtration, the solution becomes colorless. Upon rinsing the black solid with acetone, a light purple solution results.

Procedure

Please consult the Safety Information before proceeding.

1. Place 100 mL of methyl violet solution in a beaker and save for color comparison.
2. In a second beaker, place 100 mL of methyl violet solution. Add about a tablespoon of charcoal to the solution. Stir the solution thoroughly.
3. Filter the solution plus solid. Save the filtrate for color comparison.
4. Rinse the solid twice with water. Save the filtrate in a third beaker.

Safety Information

1. Methyl violet is a poison by ingestion.
2. Charcoal is a flammable solid with no toxicity data. It may contain toxic impurities.
3. Acetone is a volatile, highly flammable liquid. It should not be used near any open flames. Prolonged topical use may cause dryness, and inhalation may produce headaches.

Materials

▲   Methyl violet indicator: 0.25 g in 99.75 mL of water.

▲   Methyl violet: Dilute 10 mL of methyl violet indicator to a volume of 200 mL using water.

▲   Activated carbon (charcoal) can be purchased in an aquarium supply store.

5. Rinse the solid with acetone. Save the filtrate in a fourth beaker.

6. Compare the color of the original solution and that of the three filtrates.

Concepts

▲   Activated charcoal is used in this demonstration. Charcoal is said to be inactive when adsorbed gases cover its surface. Heating charcoal to red heat drives off the adsorbed gases and "activates" it so it may readsorb gas molecules (as in a gas mask) or adsorb organic molecules from water (as in water purification).

▲   One milliliter of finely powdered charcoal may have a total surface of 1000 m^2.

▲   Abbé Felix Fontana, physician to Duke Ferdinand II of Tuscany, was the first to devise and record an adsorption-type experiment using charcoal. He heated a piece of charcoal to redness, which removed all gases, then plunged it under mercury and held it there until it cooled. Upon inverting a tube of air over the mercury and releasing the charcoal, he found that the charcoal floated to the top of the tube, adsorbing the air. His results were published in 1777. He had previously communicated the results to Priestley about 1770, who confirmed them.

▲   The idea for this demonstration was developed by T. Lowitz in 1785.

Reactions

"Activated" charcoal adsorbs the substances that are responsible for coloration of the solution on its surface. Forces on its surface attract and hold molecules of the colored solution.

Notes

If you have access to ditto masters you can cut them up into strips and heat them in some warm water for several minutes. Dilute 10 mL to a volume of 200 mL with water and use this solution in place of the methyl violet.

References

Fowles, G. *Lecture Experiments in Chemistry;* Blakiston: Philadelphia, PA, 1937; p 95.

Kingzett, C. T. *Chemical Encyclopedia;* Van Nostrand: New York, 1928; p 111.

The Merck Index, 10th ed.; Merck: Rahway, NJ, 1983; p 250.

Nebergall, W. H.; Schmidt, F. C.; Holtzclaw, H. F., Jr. *College Chemistry;* Heath: Lexington, MA, 1976; p 656–657.

Magnetic or Not?

Yellow sulfur and gray iron solid mixed together are magnetic. When the same mixture is heated to become red hot, the resulting product is no longer magnetic.

Procedure

Please consult the Safety Information before proceeding. Perform the experiment in a well-ventilated area or a fume hood.

1. Have students record observations about the physical properties of sulfur and iron.

2. Place a mixture of 3.7 g of S and 6.4 g of Fe on a piece of paper. Show this mixture to the students. Have them record their observations.

3. Ask the students how they would separate the mixture. If a magnet is suggested, place a small amount of the mixture on a separate piece of paper and use the magnet to remove the iron.

Safety Information

1. Iron and iron sulfide may be harmful by inhalation, ingestion, or skin absorption. They may cause irritation.

2. Sulfur may cause irritation of the skin and mucous membranes.

Materials

▲ Degreased and clean iron: Submerge iron in ethanol, swirl the contents, and decant the black liquid. Rinse the filings two more times in ethanol. Rinse once in acetone. Air dry the filings.

▲ Sulfur powder.

4. Weigh a test tube. Half fill it with the mixture. Place a loose cotton plug in the end of the test tube to control the escape of sulfur vapors.

5. Heat the test tube until the mixture glows red hot (this step takes about 1 min).

6. Have a student turn some or all of the lights off so they can observe the red-hot glow. Momentarily remove the test tube from the heat so they can observe that it still glows red hot. Continue to heat the test tube for 4 more minutes.

7. Remove the tube from the heat and let the tube cool to room temperature.

8. Weigh the test tube and its contents.

9. Wrap a towel around the test tube and crack it open with a hammer.

10. Remove the contents. Show this material to the students and test it with a magnet to show that it is no longer magnetic.

Concepts

▲ A mixture contains two or more substances physically combined, each of which retains its own properties, although you may not be able to distinguish each of the different substances. The iron and sulfur form a heterogeneous mixture.

▲ When heated, the iron and sulfur chemically combine to form a compound. This compound differs in color, texture, density, solubility, and other properties from the substances of which it was made.

▲ The escaping sulfur vapors burn with a blue flame.

▲ The amounts of iron and sulfur were chosen to be stoichiometrically correct on the basis of the balanced equation. Eight moles of iron is 448 g, and 1 mol of S_8 is 256 g. Thus the mass ratio of iron to sulfur is 1.75 to 1.

▲ Mass is conserved when the mass of reactants is equal to the mass of products. Mass is seemingly not conserved in this reaction, because some sulfur vapors escape.

Reaction

$$8Fe(s) + S_8(s) \rightarrow 8FeS(s)$$

Notes

1. If the mixture is not heated to red hot for approximately 5 min, the resulting solid will still be magnetic.

2. Even though the proportion of sulfur and iron is correct, some sulfur escapes from the test tube. Decreasing the amount of sulfur does not seem to prevent this from happening.

3. Because of the escaping sulfur vapors, this demonstration should be performed in a well-ventilated area or fume hood, and the additional precaution regarding a cotton plug in the end of the test tube should be observed.

4. You may be able to demonstrate the bluish flame of burning sulfur vapors as they escape from the test tube by accidentally or intentionally igniting them.

5. Do this demonstration in a test tube that you do not mind discarding.

6. If stoichiometry has been discussed, use this demonstration to discuss why the amounts of iron and sulfur were chosen. You could write and balance the equation while the tube is cooling.

References

Benedict, F. G. *Chemical Lecture Experiments;* Macmillan: New York, 1901; pp 133–134.

Eliot, C. W.; Storer, F. H. *An Elementary Manual of Chemistry,* revised edition; Ivison, Blakeman, Taylor & Co.: New York, 1880; p 75.

Fowles, G. *Lecture Experiments in Chemistry;* Blakiston: Philadelphia, PA, 1937; pp 69 and 74.

Rate of Reaction

A black powdered alloy of sodium and lead is added to a bright red solution of universal indicator, and a gas is evolved. The solution proceeds through a spectrum of colors.

Procedure

Please consult the Safety Information before proceeding.

1. Place 50 mL of water in a beaker and add 25 drops of universal indicator.

2. Add 1 or 2 drops (or as much as needed) of 6 M HCl to make the solution bright red.

3. Add about 2 g of sodium–lead alloy to the beaker and stir the mixture briefly.

4. When the solution has changed to blue–green, add just enough HCl to obtain a bright red color again. A full color change takes about 15–20 min.

Safety Information

1. Hydrochloric acid is corrosive to all body tissues. Inhalation of the concentrated vapor may cause serious lung damage; contact with the eyes may result in a total loss of vision. Work in a fume hood, and wear gloves, a rubber apron, and goggles.

2. The sodium–lead alloy is considered nonhazardous, but prudent laboratory safety should be practiced. Inhalation of the dust may result in serious throat and lung injury. Ingestion may result in death.

3. Universal indicator is toxic by inhalation.

Materials

▲ Hydrochloric acid, 6 M HCl: Carefully add 50 mL of concentrated hydrochloric acid to enough distilled water to make 100 mL of solution.

▲ Universal indicator: commercial solution.

▲ Sodium-lead alloy, sold under various names, usually Dri-Na.

Concepts

▲ The sodium-lead alloy is 90% lead and 10% sodium. The lead inhibits the very rapid reaction of sodium with water.

▲ The sodium reacts with the water to produce hydrogen gas and sodium hydroxide.

▲ Universal indicator is blue-green in basic solution, the resulting color when sodium hydroxide is produced. The indicator is red in acid solution, which is produced by the addition of the hydrochloric acid.

Reactions

$$2Na(s) + 2H_2O(l) \rightarrow$$
$$2Na^+(aq) + 2OH^-(aq) + H_2(g)$$

Notes

1. Dri-Na is used for drying organic solvents in place of pressed sodium.

2. Any number or combination of indicators could be used to produce different color changes.

3. The amount of acid added will determine the amount of time for the initial color change but will have little effect on the time for the complete color change.

4. How long you wait after the color has changed to blue-green will determine how much acid must be added and the total number of changes that can take place. The production of sodium hydroxide continues until all of the sodium is reacted, and therefore the longer you wait to add acid, the less sodium remains. The procedure using sodium-lead alloy produces results over a span of 15–20 min. For pure sodium the time is only 15–20 s.

References

Barnard, J. J., Jr.; Johnston, M. B.; Broad, W. C. *J. Chem. Educ.* **1959**, *36*, A749.

Soroos, H. *Ind. Eng. Chem., Anal. Ed.* **1939**, *11*, 657.

Precipitating Bismuth from Pepto-Bismol

A colorless solution of sodium stannite is added to a small portion of Pepto-Bismol, producing swirls of black precipitate.

Procedure

Please consult the Safety Information before proceeding.

1. Place about 10 mL of Pepto-Bismol in a clean test tube and add 15 drops of the sodium stannite solution.
2. Immediately shake or stir the tube. Elemental bismuth will precipitate.

Concepts

▲   This classic test for the determination of bismuth ions is used to show that bismuth is con-

Safety Information

1.   Sodium hydroxide is corrosive to all tissues. Inhalation of the dust or concentrated mist may cause damage to the respiratory tract. Wear gloves, an apron, and goggles when handling it.

2.   Concentrated hydrochloric acid causes severe burns, and contact with the eyes may result in a total loss of vision. Inhalation of the vapors may cause coughing and choking. Inflammation and ulceration of the respiratory tract may occur.

Materials

▲   Pepto-Bismol: commercial product sold in drugstores.

▲   Sodium hydroxide, 6.0 M NaOH: Dissolve 12 g of NaOH in enough water to make 50 mL of solution.

▲   Sodium stannite: Add 11.5 g of $SnCl_2 \cdot 2H_2O$ to 17 mL of concentrated hydrochloric acid. Slowly add this solution to about 20 mL of water and stir. Carefully and slowly add this solution to enough water to make 100 mL of solution. Add the sodium hydroxide solution dropwise to 2 mL of the acidified tin(II) chloride solution until the precipitate that forms with the first few drops is just completely dissolved. This is a solution of sodium stannite, Na_2SnO_2.

tained in this well-known over-the-counter substance and to shed some light on the origin of the trade name.

▲   The tin in the stannite ion reduces the bismuth(III) to elemental bismuth. The bismuth(III) ion oxidizes the tin(II) in the stannite to tin(IV), producing the stannate ion.

▲   The bismuth is very finely divided and remains suspended.

Reactions

$$6OH^-(aq) + 3SnO_2^{2-}(aq) + 2Bi^{3+}(aq) \rightarrow$$
$$2Bi(s) + 3SnO_3^{2-}(aq) + 3H_2O$$

Notes

The sodium stannite must be prepared fresh, the same day you are going to use it. The precipitate that forms upon adding the sodium hydroxide is very fleeting.

References

Demming, H. G.; Arenson, S. B. *Exercises in General Chemistry and Qualitative Analysis;* Wiley: New York, 1935.

Copper Wire and Halogen Vapors

When red-hot copper wire is placed in chlorine, bromine, or iodine vapors, smoky compounds are formed.

Procedure

Please consult the Safety Information before proceeding.

1. Prepare a large test tube of iodine vapor by warming iodine crystals in a hot water bath until the inside of the test tube is filled with iodine vapor. Keep the tube submerged in the water to prevent the iodine vapor from depositing on the sides before the reaction takes place. Cover with Parafilm until used in Step 5.

2. Prepare a large test tube of bromine vapor by warming pyridinium bromide perbromide in a hot water bath. Cover with Parafilm until used in Step 5.

Safety Information

1. Concentrated hydrochloric acid causes severe burns, and contact with the eyes may result in a total loss of vision. Inhalation of the vapors may cause coughing and choking. Inflammation and ulceration of the respiratory tract may occur.

2. For work with all elemental halogens, work in a fume hood and wear goggles, gloves, and a rubber apron. Concentrated halogen vapor causes edema of the respiratory tract, spasm of the glottis, and asphyxia. →

3. Commercial bleach can causes severe burns, and contact with the eyes may result in a total loss of vision. Inhalation of the vapors may cause coughing and choking. Inflammation and ulceration of the respiratory tract may occur.

4. Pyridinium bromide perbromide is a safer method of producing bromine water than using sulfuric acid and sodium bromide or vials of liquid bromine; however, extreme care must be exercised. Contact with the liquid will result in tissue damage. Inhalation of the vapor may cause serious lung damage; contact with the eyes may result in a total loss of vision. Work in a fume hood and wear gloves, a rubber apron, and goggles.

3. Prepare a large test tube of chlorine vapor by adding 4 or 5 drops of concentrated HCl to about 5 mL of 5% sodium hypochlorite (bleach). Cover with Parafilm until you are ready for Step 5.

4. Heat the coiled copper wire in a Bunsen burner flame until red hot.

5. Place the red-hot coil into the chlorine vapor and observe the dense cloud of $CuCl_2$ that forms around the coil and inside the tube. The coil will also continue to glow red hot.

6. Repeat Steps 4 and 5 for the bromine vapor.

7. Repeat Steps 4 and 5 for the iodine vapor.

8. Compare the reactions on the basis of the amount of halide formed and the glow of the coil.

Concepts

▲   The hot copper and the chlorine vapor have enough energy to react. The heat of the reaction also keeps the copper wire glowing. Some copper chloride is formed on the wire and produces a copper flame test when you reheat the coil in the burner flame.

▲   The reactivity of the halogens is demonstrated very well: the intensity of the afterglow, the quantity of the smoke (powdered copper halides), and the intensity of the flame test—corresponding to the intensity of attack by the halogen vapor.

▲   The three halogens are produced by different methods. The iodine is a solid that is heated, and the vapor attacks the elemental copper very slowly and to a smaller extent than do those of bromine and chlorine. The bromine is slightly soluble as dissolved molecular bromine, which upon warming degases from the solution and attacks the elemental copper, producing a more visible reaction and more visible product than those of the iodine. The chlorine is bound to oxygen and dissolved in the water. Upon the addition of hydrogen ions, the hypochlorite ion produces nascent chlorine and then molecular chlorine, which degases quickly and attacks the elemental copper, producing a copious amount of copper chloride "smoke". (You will recall that smoke is an aerosol, a solid dispersed in a gas.) Copper(II) chloride is formed, but copper(I) bromide and copper(I) iodide are formed.

Reactions

$$2H^+(aq) + Cl^-(aq) + OCl^-(aq) \rightarrow H_2O(l) + Cl_2(g)$$

$$2C_5H_6NBr \cdot Br_2(s) + H_2O(l) \rightarrow$$
$$2C_5H_6NBr\ (aq) + Br_2(aq) \xrightarrow{\Delta} Br_2(g)$$

$$I_2(s) \xrightarrow{\Delta} I_2(g)$$

$$Cu(s) + Cl_2(g) \rightarrow CuCl_2(s)$$

$$2Cu(s) + Br_2(g) \rightarrow 2CuBr(s)$$

$$2Cu(s) + I_2(g) \rightarrow 2CuI(s)$$

Materials

▲ Coiled copper wire: Make a coil by wrapping No. 12 gauge wire around a pencil. The coil should be about 2 cm long, and the individual turns should be spaced slightly apart.

▲ Hydrochloric acid, concentrated HCl: stock solution.

▲ Sodium hypochlorite, 5% NaOCl: commercial bleach. Alternatively, NaOCl solution can be prepared by adding 5 g of solid NaOCl to 95 mL of distilled water.

▲ Pyridinium bromide perbromide: Place about 1 g of pyridinium bromide perbromide in 10 mL of water in a test tube.

▲ Iodine crystals: Place a matchhead-sized amount of iodine crystals in a test tube.

References

Garman, R. P. *J. Chem. Educ.* **1969**, *46*, A310.

Merker, P. C.; Vona, J. A. *J. Chem Educ.* **1949**, *26*, 613–614.

Partington, J. R. *A Textbook of Inorganic Chemistry;* Macmillan: London, 1950; p 725.

Rapid Oxidation of the Iron(II) Ion

A purple solution of potassium permanganate containing a suspended piece of steel wool will be spontaneously "cleaned up" to colorless in about 1 min.

Procedures

Please consult the Safety Information before proceeding.

1. Add about 400 mL of water to a 600-mL tall-form beaker. Add a few grains of potassium permanganate to the beaker and acidify the mixture with 5 drops of concentrated sulfuric acid. Place the beaker on a magnetic stirrer and stir the mixture at medium speed.

2. Suspend a golfball-sized piece of steel wool from the top of the beaker into the solution. Continue to stir the mixture. Observe.

3. Sulfuric acid is corrosive to all body tissues. Inhalation of the concentrated vapor may cause serious lung damage; contact with the eyes may result in a total loss of vision. Work in a fume hood and wear gloves, a rubber apron, and goggles.

Concepts

▲ Steel wool is a very fine iron ribbon, and the iron reacts with the slightly acidic solution to produce iron(II) ions. The iron(II) is easily oxidized to the iron(III) state by the permanganate ion.

▲ The iron(II) ion reduces the manganese(VII) to manganese(II) as it is oxidized. As the permanganate is reduced, the color fades from purple to colorless.

Reactions

1. $Fe(s) + 2H^+(aq) \rightarrow Fe^{2+}(aq) + H_2(g)$

2. $5Fe^{2+}(aq) + MnO_4^-(aq) + 8H^+(aq) \rightarrow$
$$5Fe^{3+}(aq) + Mn^{2+}(aq) + 4H_2O$$

Notes

1.   Add just enough $KMnO_4$ to color the solution, not to darken it.

2.   Be sure the steel wool is clean. Do not use the soap-filled variety. It is helpful to clean the steel wool in acetone before use.

Materials

▲ Sulfuric acid, concentrated H_2SO_4: stock solution.

▲ Potassium permanganate crystals, $KMnO_4$.

▲ Steel wool from a hardware store.

References

Demming, H. G.; Arenson, S. B. *Exercises in General Chemistry and Qualitative Analysis;* Wiley: New York, 1935.

Reaction of Two Elements by Direct Combination

When gallium metal and iodine crystals are ground together in a mortar and pestle, a yellowish cloud of smoke is produced. The gallium and the iodine reacted on contact, releasing heat and producing a fine dust of gallium iodide that is carried upward by the heated air.

Procedure

Please consult the Safety Information before proceeding.

1. If the gallium is not in the liquid state, place a pea-sized piece in a test tube and warm it to 40 °C in a water bath.

2. After the gallium is melted, place the small drop of gallium metal in a **clean** mortar and add an equal-sized amount of iodine crystals.

3. Grind the two together. After a few seconds, a hissing sound will be produced with the release

2. Iodine solid and vapor are both harmful to the skin and especially the respiratory tract. Contact with the solid will result in tissue damage. Inhalation of the vapor may cause serious lung damage; contact with the eyes may result in a total loss of vision.

of a cloud of GaI_3 dust. A soft deep red compound remains behind in the mortar. Dissolve this solid in water to form a deep red solution. A light yellow powder will remain on the upper edges of the mortar.

Concepts

▲ When the gallium metal is ground together with the iodine crystals, the pure elements react.

▲ The reaction appears to be exothermic with the release of enough heat energy to melt the iodine crystals.

▲ Several authors have shown that GaI_3 reacts with excess Ga metal to produce red $Ga(GaI_4)$ and that other gallium trihalides react similarly.

▲ The red solid that remains is probably $Ga(GaI_4)$. It dissolves in water to form a deep red solution. This solution is probably $GaI–GaI_3$; the gallium is not Ga(II) but rather Ga(I) and Ga(III). Equimolar amounts of GaI and GaI_3 produce the $Ga(GaI_4)$.

Reactions

$$2Ga(l) + 3I_2(s) \rightarrow 2GaI_3(s)$$

　　silver　　silver-gray　　　yellow

$$2Ga(l) + I_2(s) + 2GaI_3(s) \rightarrow 2Ga(GaI_4)(s)$$

silver　　silver-gray　　yellow　　　　　　red

Notes

1. This reaction is similar to the mercury and iodine reaction noted by George Fowles in 1937. However, mercury vapor is so dangerous that you probably should not use that demonstration to show the reactivity of elements.

2. The gallium solid can be placed directly in the mortar and ground up. The heat of friction will melt the gallium, and the melting step can be eliminated. An aerosol (smoke) is a solid suspended in a gas, and thus the product is not a vapor but rather smoke.

3. Professor Hubert Alyea related a story concerning the familiar mercury and iodine reaction. Alyea sent lecture notes and demonstration notes to George Kistiakowski, who was teaching freshman chemistry at Smith College at the time. Kistiakowski apparently had not read the demonstration notes carefully and had only read the direction to grind mercury and iodine with a mortar and pestle. He apparently put in a very large amount of each and proceeded to grind away in the front of the lecture hall. In a few seconds the reaction occurred with the release of a loud noise and large cloud of purple iodine vapor. The professor's face, hands, and clothing were covered with a purple residue, after which he promptly dismissed the class.

Materials

▲ Gallium metal.
▲ Iodine crystals.

References

Beamish, J. C.; Small, R. W. H.; Worral, I. J. *Inorg. Chem.* **1979**, *18*, 220–223.

Benedict, F. G. *Chemistry Lecture Experiments;* Macmillan: New York, 1901; pp 381–382.

CRC Handbook of Chemistry and Physics, 66th ed.; Weast, R. C., Ed.; CRC Press: Boca Raton, FL, 1985.

Fowles, G. *Lecture Experiments in Chemistry;* Blakiston: Philadelphia, PA, 1937.

Greenwood, N. N.; Earnshaw, A. *Chemistry of the Elements;* Pergamon: New York, 1984; pp 266–272.

The Merck Index, 10th ed.; Merck: Rahway, NJ, 1983; p 250.

Newth, G. S. *Chemical Lecture Experiments;* Longman, Green, & Co.: New York, 1892; p 106.

Gold Dust and Nuggets from Copper

When granular copper or shot is placed in a strong alkali solution with granular zinc and then heated, "gold" is produced.

Procedure

Please consult the Safety Information before proceeding.

1. Place 50 mL of saturated potassium hydroxide solution in a 150-mL beaker. Add a spoonful of granular zinc. Stir carefully.

2. Place 10 or 15 copper nuggets in the solution and let the mixture stand for 2–3 h.

3. When the nuggets are completely covered with a silver sheen, remove and wash them thoroughly with distilled water.

4. Pick up a nugget with metal forceps and heat it carefully in the blue portion of a Bunsen burner flame until a gold-colored sheen is produced.

Safety Information

Potassium hydroxide is corrosive to all tissues. Inhalation of the dust or concentrated mist may cause damage to the respiratory tract. Wear gloves, an apron, and goggles when handling.

Materials

▲ Potassium hydroxide, KOH saturated: Dissolve 50 g in 50 mL of water.

▲ Zinc granules, 150 mesh.

▲ Copper shot.

▲ Zinc electrode.

Variation

1. Place a spatulaful of fine granular copper in a large test tube and add 25 mL of saturated potassium hydroxide solution. Place a cleaned zinc electrode in the test tube. Let it remain in the test tube overnight.

2. When the copper granules are silver-colored, remove the zinc electrode and decant the potassium hydroxide solution back into the original beaker. Wash the copper granules thoroughly with distilled water until the water is no longer basic to a drop of phenolphthalein. Spread the granules on a piece of filter paper. Let them dry thoroughly.

3. Place half of the silver-colored granules in a crucible and heat them until gold-colored. The remaining half can be used as "silver" dust.

Concepts

▲ The zinc plates out on the copper by reactions shown below. Potassium hydroxide aids in the plating action. This reaction is the reverse of what we normally expect the spontaneous electrochemical reaction to be. Zn metal is the reducing agent and the driving force.

▲ The heating of the copper and zinc coating causes a type of brass to be formed. Although the temperature reached in the Bunsen burner flame is not sufficient to melt copper, there is mixing of the two metals.

Reactions

1. $Zn(s) + 2OH^-(aq) + 2H_2O \rightarrow Zn(OH)_4^{2-}(aq) + H_2(g)$
   $E_{rxn} = +0.4$ V

2. $Zn(OH)_4^{2-} + 2e^- \rightarrow Zn + 4OH^-$ $E^\circ = -1.23$ V

3. $2H_2O + 2e^- \rightarrow H_2 + 2OH^-$ $E^\circ = -0.83$ V

4. $Cu(s) + Zn(s) \overset{\Delta}{\rightarrow}$ brass

Notes

1. This experiment is a new variation of the old "gold penny" reaction. If you use copper shot of an irregular shape, wonderful "gold" nuggets appear.

2. If you try to use granular copper with the granular zinc, they cannot be separated after the zinc-plating process. If a strip of zinc is placed in the potassium hydroxide solution, the zinc is present but in an easily removable form.

References

Summerlin, L.; Ealy, James L., Jr. *Chemical Demonstrations: A Sourcebook for Teachers;* American Chemical Society: Washington, DC, 1985; Vol. 1, p 104.

Szczepankiewicz, S. H.; Bieron, T. F.; and Kozik, M. *J. Chem. Ed.* **1995,** *72,* 386–389.

Corrosion, Complex Ions, and Passivity

A blue solution is added to a test tube containing an iron nail resting on top of solid sodium chloride. Immediately, many colorful reactions take place. In about 1 day, the nail will be covered in rust or may have disappeared.

Procedures

Please consult the Safety Information before proceeding.

1. Add 3 g (1/3 tsp) of solid sodium chloride to each of three test tubes.

2. Push a piece of filter paper, slightly larger than the inside diameter of the tube, into each tube down onto the salt.

3. Add 10 mL of copper sulfate solution to each test tube. Record your observations.

4. Place one "bright" 2-penny nail in concentrated nitric acid for 30 s. Place a second nail in 30% hydrogen peroxide for 30 s. Do not treat the third nail in any way.

5. Place one nail in each test tube, noting which nail is in each. Observe all reactions and record.

Concepts

▲   When the copper sulfate reacts with the solid sodium chloride, yellow tetrachlorocuprate(II) anions, $CuCl_4^{2-}$, are formed. Migration of Cu^{2+} through the filter paper to the Cl^- ion allows this complex to form. Although the tetrachlorocuprate(II) anions in aqueous solution are yellow, when they mix with the copper(II) ions, which are blue, the resulting color that we observe is green.

▲   Chloride ions aggressively aid the corrosion of the iron, and they help in charge transfer between the microelectrochemical cells that form at stress points on the surface of the nail. When wire nails are extruded, they are placed under tremendous stress at the head and point. Consequently, these are the first sites of corrosion and chemical activity.

▲   The nail placed in the concentrated nitric acid does not react with the copper in the test tube. Even with the presence of chloride ions, the nail is passive. The nitric acid causes the formation of a thin coating of FeO on the surface and prevents the reaction with the copper ions. The nail placed in 30% hydrogen peroxide is also rendered passive and fails to react for the same chemical reason.

▲   Some nails are coated with zinc. Nitric acid also reacts with the zinc coating and oxidizes it to white zinc oxide, thus preventing the nail from reacting with the copper ions.

▲   If the nail is not made passive with an oxidizer, the zinc reacts with the copper ions until the thin coating is depleted, resulting in the formation of zinc ions and copper metal.

▲   *The electrochemical reaction:* An iron nail placed in a copper sulfate solution results in the plating out of copper metal on the nail.

Copper is below iron in the electrochemical series, iron metal is oxidized to Fe^{2+} ions through loss of electrons, and Cu^{2+} ions are reduced to copper atoms by gaining electrons.

▲ There will be evidence of red–brown iron(III) oxide and off-white iron(II) hydroxide formed on the nail as it corrodes. In the presence of water and some oxygen, corrosion is much faster than with little or no oxygen. If plentiful oxygen is present, a black coating of iron(II) and iron(III) oxides is formed.

Materials

▲ Copper sulfate pentahydrate, 1.0 M $CuSO_4 \cdot 5H_2O$: Dissolve 25 g of copper sulfate in enough water to make 100 mL of solution.

▲ Nitric acid, concentrated HNO_3: stock solution.

▲ Hydrogen peroxide, 30% H_2O_2: stock solution.

▲ Sodium chloride.

▲ Three "bright" 2-penny nails.

Reactions

1. $Cu^{2+}(aq) + 4Cl^-(aq) \rightarrow CuCl_4^{2-}(aq)$

2. $2Fe(s) \rightarrow 2Fe^{2+}(aq) + 4e^-$

3. $8e^- + 2O_2(g) + 4H_2O \rightarrow 8OH^-(aq)$

4. $2Fe(s) + O_2(g) + 2H_2O \rightarrow 2Fe(OH)_2(s)$

5. $6Fe(OH)_2(s) + O_2(g) \rightarrow 2Fe_3O_4 \cdot H_2O(s) + 4H_2O(l)$

6. $Zn(s) + Cu^{2+}(aq) \rightarrow Cu(s) + Zn^{2+}(aq)$

7. $2Fe(s) + 3Cu^{2+}(aq) \rightarrow 2Fe^{3+}(aq) + 3Cu(s)$

Notes

1. This demonstration is taken from an Indian patent for an electrochemical cell. In the patent, solid copper sulfate crystals and solid sodium chloride were used with an iron mesh between the two layers.

2. This experiment makes an excellent observation demonstration for the first day of class.

3. We have used several different solids—alkali and alkaline earth halides. Although these combinations give very interesting results, no combination seems to work as well as sodium chloride and copper sulfate. However, aluminum, zinc, magnesium, or steel wool works well for the metal.

References

Drozdov, J. *Appl. Chem. USSR (Engl. Transl.)* **1958**, *31*, 202.

Kotz, J. C.; Purcell, K. F. *Chemistry and Chemical Reactions;* Saunders: Philadelphia, PA, 1987.

Mathur, P. B. *J. Chem. Educ.* **1962**, *39*, A897.

Nebergall, W. H.; Schmidt, F. C.; Holtzclaw, H. F., Jr. *College Chemistry*, 5th ed.; Heath: Lexington, MA, 1976; p 871.

Enthalpic Conversion of Bismuth Iodide to Bismuth Iodite

When a deep red solution is added to a colorless solution, a black precipitate is produced. If water containing this precipitate is heated to boiling, a red solution results.

Procedure

Please consult the Safety Information before proceeding.

1. Place 5 mL of bismuth nitrate solution in a test tube. Add 10 drops of potassium iodide–iodine solution. Observe the dense black precipitate.

2. Let the mixture sit for several minutes or centrifuge it. Decant the solution, wash the precipitate several times, and discard the supernatant liquids.

3. Add 5 mL of distilled water to the precipitate and heat the mixture to boiling. Observe the

3. For iodine and all elemental halogens, work in a fume hood and wear goggles, gloves, and a rubber apron. Concentrated halogen vapor causes edema of the respiratory tract, spasm of the glottis, and asphyxia.

color change from black to red and the dissolving of the precipitate.

Concepts

▲ Bismuth(III) ions react with iodide ions to produce a dense black precipitate of bismuth triiodide. When this precipitate is heated in boiling water, bismuth iodite is formed. The change from an insoluble black precipitate to a soluble compound forming a red solution is dramatic and indicative of a chemical change.

▲ The reason for the change from iodide to iodite is not clear. Bismuth iodite is a brick-red insoluble compound that decomposes above 300 °C. Some other reactions are taking place that may involve the polyiodide ion. Simple heating of the dry bismuth iodide in air will also convert it to the brick-red bismuth iodite.

Materials

▲ Bismuth nitrate, 0.1 M $Bi(NO_3)_3 \cdot 5H_2O$: Dissolve 4.8 g in enough water to make 90 mL of solution. Add concentrated nitric acid dropwise, stirring after each drop until the solution clears. Add an additional drop of acid. Then add enough water to make 100 mL of solution.

▲ Potassium iodide-iodine, 0.1 M $KI–I_2$: Dissolve 1.6 g of potassium iodide in enough water to make 100 mL of solution. Add 0.5 g of iodine crystals and mix the solution thoroughly.

Reactions

$$Bi^{3+}(aq) + 3I_3^-(aq) \rightarrow BiI_3(s) + 3I_2(s)$$

$$\underset{\text{black}}{BiI_3(s)} + H_2O(l) \rightarrow \underset{\text{red}}{BiOI(s)} + 2HI(aq)$$

Notes

1. This reaction will not work with potassium iodide alone, only with $KI–I_2$.

2. The precipitate must be washed several times with distilled water before it will change to red bismuth iodite when heated.

References

Feigl, F. *Spot Tests, Vol. I, Inorganic Chemistry*, 4th English translation; Elsevier: Houston, TX, 1954; p 76.

Greenwood, N. N.; Earnshaw, A. *Chemistry of the Elements*; Pergamon: New York, 1984; pp 667, 979–982.

Reduction of Silver with Bluing

When silver nitrate is added to laundry bluing and made slightly basic with aqueous ammonia, a silver mirror appears.

Procedure

Please consult the Safety Information before proceeding.

1. Add 5 mL of laundry bluing to a large test tube. Add an equal amount of silver nitrate solution. Stopper and shake the tube to mix thoroughly.

2. Let the mixture sit for 1 h. During this time the suspended matter will darken to an intense black and begin to precipitate.

3. Prepare a large test tube for mirroring. Roughen the inside by scratching with steel wool. Rinse the tube with distilled water and then rinse with a few milliliters of concentrated aqueous ammonia. Use in Step 4.

Materials

▲ Bluing: Commercial laundry product purchased from the grocery store. Be sure it contains the compound ultramarine.

▲ Silver nitrate, 0.1 M $AgNO_3$: Dissolve 1.7 g in enough water to make 100 mL of solution. Store in an amber bottle.

▲ Aqueous ammonia, concentrated NH_3: Use stock solution.

4. Centrifuge and decant the liquid from Step 2 into the test tube prepared in Step 3. Let the liquid sit for about an hour. A silver mirror will form on the inside of the test tube up to the level of the liquid.

Reactions

$$Na_8S_2(Al_6Si_6O_{24}) + 3AgNO_3(aq) + NH_3(aq) \rightarrow$$
$$Ag(s) + Ag_2S + \text{complex silicates}$$

Concepts

▲ Ultramarine, the ingredient in bluing, is a framework silicate that has silicon and aluminum atoms at the corners of a polyhedron. Sodalite, the naturally occurring mineral, is white with the formula $Na_8Cl_2Al_6Si_6O_{24}$, with chloride ions in place of sulfides that would be found in ultramarine. As the chloride ions are partially replaced with sulfide ions, the color becomes the brilliant blue of the mineral lapis lazuli.

▲ The color of an object is the result of the interaction of electrons of the surface particles with photons of ultraviolet and visible light. Some photons are absorbed and others are reflected, depending upon the nature of the dye or pigment. In ultramarine, the S_2^- and the S_3^- radicals absorb most of the visible spectrum and reflect the intense blue photons associated with the color ultramarine. Therefore, the "dingy yellow" wavelengths associated with laundry are absorbed and cannot be reflected. The ultraviolet wavelengths from the sun are reflected and add to the "whitening" of the laundry.

▲ The reaction appears to be similar to that of the standard silver mirror. That is, the silver is reduced from the 1+ cation to the neutral atom by the addition of electrons. The bluing agent, ultramarine adsorbed on methylcellulose in suspension, becomes black. This color change most likely is a reaction between the sulfide in the ultramarine and the silver cations to form insoluble black silver sulfide.

▲ The reducing agent in this reaction is not clear. The S_2^- and S_3^- are electron-deficient and are probably not acting as a reducing agent in this reaction.

▲ This demonstration is an excellent example of *electroless* plating as described by Justus von Liebig in 1835. The main difference between electrode and electroless plating is the absence of electrodes. Also, reducing agents such as aldehydes and glyoxal supply the electrons in solution. The surface of the substance must be scrupulously cleaned and most often needs to be prepared with a catalytic substrate. In this case, the reducing agent is not certain, and the surface is not treated with a catalyst. However, the surface is cleaned and scratched to provide a high surface area for the reducing agent to interact with the silver ions.

Notes

1. This reaction was discovered serendipitously. We had read that the sodium in ultramarine could be replaced by potassium or silver ions. We found no further information about the state of the ultramarine or silver; thus, we tried various solutions. According to Kingzett (1928), silver replacement would produce a yellow ultramarine. A 0.1 M silver nitrate solution was added to several milliliters of bluing. Upon standing for several hours, the solution did not become yellow, but a silver mirror formed. Another test tube was used, and again a silver mirror formed. At this point, we suspected that something novel had been produced. However, all further attempts to reproduce the mirror with just bluing and silver nitrate failed. We suspected that the new bottle of bluing had settled somewhat. The organic compounds added by the manufacturer may have become concentrated in the solution at the top of the bottle. These organic compounds may have acted as reducing agents, promoting the plating of silver. This behavior may explain why the reaction originally worked without the addition of aqueous ammonia and subsequently did not.

2. The reaction mixture prepared in Step 2 does not need to be centrifuged; it will precipitate slowly, and with the addition of a few drops of concentrated aqueous ammonia, a silver mirror will form. The mirror produced in this manner may not be as complete because of the immediate presence of the flocculant material.

References

Dermott, J. *Electroless Plating;* Noyes Data: Park Ridge, IL, 1972.

Ealy, Julie B. *Sci. Teach.* **1991**, *58(4)*, 26–27.

Ealy, Julie B.; Ealy, James L., Jr. *Close-up on Chemistry;* American Chemical Society: Washington, DC, 1991.

Feigl, F. *Spot Tests, Vol. I, Inorganic Chemistry*, 4th English translation; Elsevier: Houston, TX, 1954; p 287.

Greenwood, N. N.; Earnshaw, A. *Chemistry of the Elements;* Pergamon: New York, 1984; pp 416, 921.

Kingzett, C. T. *Chemical Encyclopedia;* Van Nostrand: New York, 1928; p 111.

Kirk–Othmer Encyclopedia of Chemical Technology, 3rd ed.; Grayson, M.; Eckroth, D., Eds.; Wiley: New York, 1979; Vol. 8.

The Merck Index, 10th ed.; Merck: Rahway, NJ, 1983; p 250.

Partington, J. R. *A Textbook of Inorganic Chemistry;* McMillan: London, 1950; p 935.

Reaction of Carbon and Iodate Ion

When carbon is added to potassium iodate and the mixture is heated, a spectacular reaction produces free iodine vapor and potassium iodide.

Procedure

Please consult the Safety Information before proceeding. Perform this demonstration in a fume hood or in a well-ventilated room.

1. Add 1 g of finely powdered potassium iodate to each of three small Pyrex test tubes. Add 0.05 g of powdered charcoal or carbon to the first test tube. Mix the contents thoroughly with a stirring rod.

2. Heat the first test tube containing potassium iodate and charcoal gently in the flame of a Bunsen burner until the contents begin to glow. Remove the tube from the heat and observe

Materials

▲ Potassium iodate powder.

▲ Carbon, amorphous powder or activated powder (see Notes).

▲ Graphite, powder (see Notes).

the continuing reaction and release of iodine vapor.

3. Gently heat the second test tube containing potassium iodate and observe that the solid does not decompose.

4. Place powdered graphite in the third test tube of potassium iodate. Heat the tube gently and observe that no reaction takes place.

Concepts

▲ When potassium iodate is heated to less than 535 °C, it will not decompose. However, when it is heated in the presence of amorphous carbon, the iodate is reduced by the carbon. The reaction releases energy in the form of heat, and this heat sustains the reaction after it is removed from the flame.

▲ In graphite, the carbon is bonded in a structural form that does not allow iodate to be reduced by it, so no visible reaction takes place. Iodate may be reduced by graphite, but the reaction rate is very slow at temperatures below 500 °C. The reaction rate appears to be similar to the rate of the reaction when pure potassium iodate is heated.

▲ The reaction, using amorphous carbon rather than graphite, was employed until recently as a forensic test to determine if an unknown substance contained vegetable or animal matter. The pathologist would char the unknown material and place a small amount of the ash in a test tube with potassium iodate. If a reaction occurred, the unknown substance contained organic matter.

Reactions

1. $3KIO_3(s) + 4C(\text{amorphous}) \rightarrow$
$$KI(s) + 3CO_2(g) + I_2(g) + K_2CO_3(s)$$

2. $KIO_3(s) + C(\text{graphite}) \rightarrow$
$$\text{no reaction (or very slow)}$$

Notes

▲ Powdered graphite can be produced by breaking off the point of a sharpened pencil and grinding it in a mortar and pestle. Amorphous carbon can be soot from a candle or finely powdered activated charcoal. Activated charcoal is easily obtained as aquarium filter charcoal.

References

Feigl, F. *Spot Tests, Vol. I, Inorganic Chemistry*, 4th English translation; Elsevier: Houston, TX, 1954; p 345.

Reactions Involving Gases

Sodium Hydrogen Carbonate Turns Red Indicator Yellow

Pouring a colorless sodium hydrogen carbonate solution into a red acidic solution containing methyl red indicator and liquid detergent results in a yellow foamy solution resembling beer.

Procedure

Please consult the Safety Information before proceeding.

1. Place the red acidic solution in a 400-mL beaker. Add a squirt of liquid detergent and gently mix the two together.

2. When ready, pour the sodium hydrogen carbonate solution into the 400-mL beaker. Foaming over might occur.

Safety Information

1. External contact with hydrochloric acid causes severe burns, and contact with the eyes may result in a total loss of vision. Inhalation causes coughing and choking with possible inflammation and ulceration of the respiratory tract. Work in a fume hood and wear goggles, gloves, and a rubber apron.

2. Sodium hydrogen carbonate is moderately toxic by ingestion.

Materials

▲ Sodium hydrogen carbonate, $NaHCO_3$: Dissolve 10 g in 150 mL of water.

▲ Hydrochloric acid, 6.0 M HCl: Adding concentrated acid to water, mix 50 mL of acid with enough distilled water to make 100 mL of solution.

▲ Methyl red: Dissolve 0.02 g in 60 mL of ethanol plus 40 mL of water.

▲ Red acid solution: Mix 8 mL of the 6.0 M HCl with 150 mL of water. Add enough methyl red indicator to make a rose-pink color.

▲ Liquid detergent.

Concepts

▲ When acid reacts with sodium hydrogen carbonate, carbon dioxide gas is evolved and water is formed. Adding detergent provides a suitable phase for formation of foam as carbon dioxide is evolved. Without detergent, the foam produced is very short-lived.

▲ Methyl red indicator changes in the pH range 4.4–6.2. It is red in acid and yellow in basic solution. When the acidic solution and sodium hydrogen carbonate are mixed, the loss of H_3O^+ ions to form water results in a more basic solution. The indicator becomes yellow.

▲ When the concentration of hydroxide ion, OH^-, increases during the reaction, the equilibrium shifts toward the product side of the indicator reaction, where methyl red is yellow.

▲ Beer contains proteins and carbohydrates that promote foaming. These concentrate at the surface of the beer, altering the forces of attraction at the surface and preventing interaction of the beer with the proteins and carbohydrates. When there is any agitation or splashing, such as in pouring the beer, fresh surface is exposed, which is immediately replaced with the proteins and carbohydrates that interact with the beer. The fresh surface containing carbon dioxide remains on the outside of the beer, and the gas is supported inside and out by the proteins and carbohydrates. Foaming appears in the beer at this point.

Reactions

$$HCO_3^-(aq) + H_3O^+(aq) \leftrightharpoons 2H_2O(l) + CO_2(g)$$

$$\underset{\text{red}}{HIn(aq)} + OH^-(aq) \leftrightharpoons H_2O(l) + \underset{\text{yellow}}{In^-(aq)}$$

where In is the symbol for the indicator.

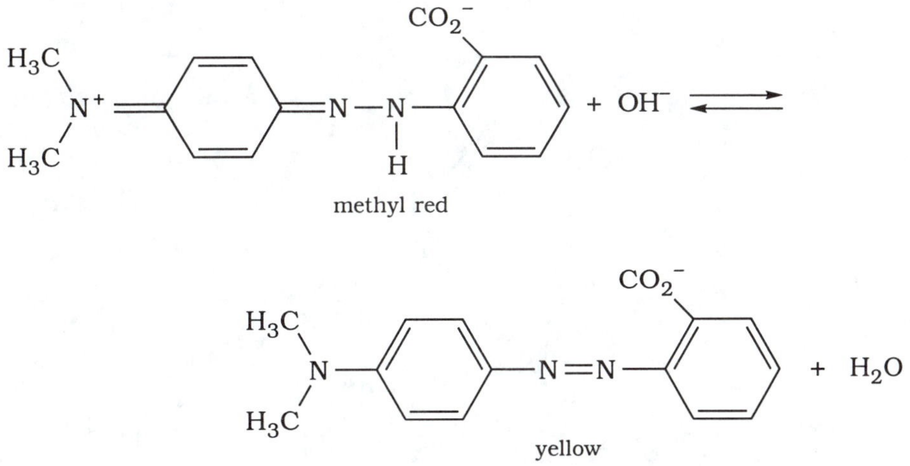

methyl red

yellow

Notes

▲ A 600-mL tall-form beaker prevents any foaming over.

▲ The demonstration works without adding detergent, but the foam is much more realistic looking and longer lasting with detergent.

References

Lippy, J. D., Jr.; Palder, E. L. *Modern Chemical Magic;* Stackpole: Harrisburg, PA, 1959; p 5.

Miller, G. T.; Lygre, D. G.; Smith, W. D. *Chemistry, A Contemporary Approach,* 2nd ed.; Wadsworth: Belmont, CA, 1987; pp 381–384.

Nebergall, W. H.; Schmidt, F. C.; Holtzclaw, H. F., Jr. *College Chemistry,* 5th ed.; Heath: Lexington, MA, 1976; p 944.

The Production of Two Foams

Sodium hydrogen carbonate and tartaric acid solids are mixed in a tall cylinder. Upon addition of a soapy solution, foam begins to move up through the cylinder.

Procedure

Please consult the Safety Information before proceeding.

1. In a 1-L cylinder, place 7 g of sodium hydrogen carbonate and 7 g of tartaric acid. Mix the contents together with a long stirring rod.

2. Add half of the soapy solution. Do not mix.

3. Observe the slow production and movement of foam.

4. In a second 1-L cylinder, place six crushed tablets of Alka-Seltzer.

Materials

▲ Soapy solution: Mix 50 mL of liquid detergent with 400 mL of water.

▲ Sodium hydrogen carbonate powder.

▲ Tartaric acid powder.

▲ Alka-Seltzer: commercial product available in drug stores.

5. Add the other half of the soapy solution. Do not mix.

6. Again, observe the slow production and movement of foam.

Concepts

▲ When solutions of sodium hydrogen carbonate and tartaric acid are mixed, carbon dioxide, CO_2, gas is evolved. Alka-Seltzer contains sodium hydrogen carbonate and an acid component, citric acid.

▲ Hydrogen on the hydroxyl group of tartaric acid is replaced by sodium. A similar reaction occurs with Alka-Seltzer.

▲ The soapy solution provides a suitable phase for formation of foam as gas is evolved.

▲ Sodium dodecyl sulfate, $C_{12}H_{25}OSO_3Na$, an alkyl sulfate, is known for its cleaning ability and voluminous, stable foam. Alkyl sulfates derived from linear alcohols containing 12–15 carbons are used in products such as carpet shampoos and light-duty household cleaners. When 16–18 carbons are present, they are used in heavy-duty household detergents.

▲ The preferred fire-extinguishing agent for fighting large-scale chemical or petroleum fires is a protein-based material that is mixed with water and aerated at the nozzle. The foam produced covers the flaming surface, excludes air, and smothers the fire.

Reactions

Notes

1. For the production of other foams, consult the References.

2. Try various liquid detergents to determine which work best.

3. Have students time how long it takes for the foam to reach the top using different detergents.

4. Crush the Alka-Seltzer tablets in the packet they come in using a heavy object such as a hammer.

References

Alexander, A. E.; Johnson, P. *Colloid Science;* Clarendon: Oxford, England, 1949 (foam).

Alyea, H. N. *J. Chem. Educ.* **1956**, *33*, A505 (foam).

Birkerman, J. J.; Perri, J. M.; Booth, R. B.; Currie, C. C. *Foams: Theory and Industrial Applications;* Reinhold: New York, 1953.

Ealy, Julie B.; Ealy, James L., Jr. *Close-Up on Chemistry: Chemical Demonstrations;* American Chemical Society: Washington, DC, 1991; pp 23–24 (foam).

Fowles, G. *Lecture Experiments in Chemistry;* Blakiston: Philadelphia, PA, 1937; p 23.

George, B. *J. College Sci. Teaching* **1986**, *16*, 75 (foam).

Kemp, D. S.; Vellaccio, F. *Organic Chemistry;* Worth: New York, 1980; pp 335.

Russ, S. In *Kirk-Othmer Encyclopedia of Chemical Technology*, 3rd ed.; Grayson, M.; Eckroth, D., Eds.; Wiley-Interscience: New York, 1980; Vol. II, pp 141–143.

Basic Cabbage Juice and Carbon Dioxide

A green basic cabbage juice solution becomes a light lavender shade when carbon dioxide gas is bubbled into the solution.

Procedure

Please consult the Safety Information before proceeding.

Set up the apparatus as illustrated on the following page and proceed.

1. Place 10–15 g (2–3 tsp) of calcium carbonate in a 250-mL flask fitted with a two-hole stopper. One hole contains a straight piece of glass tubing to deliver hydrochloric acid into the flask. The other hole contains a glass bend tube with a piece of rubber tubing attached that leads into a test tube containing basic cabbage juice.

Safety Information

1. Calcium carbonate is a skin and eye irritant and is slightly toxic by ingestion.

2. External contact with hydrochloric acid causes severe burns, and contact with the eyes may cause permanent damage. Inhalation causes coughing and choking with possible inflammation and ulceration of the respiratory tract. Work in a fume hood and wear goggles, gloves, and a rubber apron.

3. Sodium hydroxide is corrosive to all tissues. Inhalation of the dust or concentrated mist may cause damage to the respiratory tract. Wear gloves, an apron, and goggles when handling.

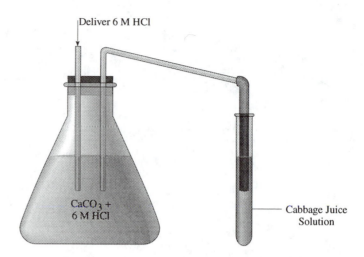

Materials

▲   Cabbage juice: Cut a red cabbage into small pieces and boil it in distilled water for 1 h. Pour off the purple solution and use it for the demonstration.

▲   Basic cabbage juice: Place 50 mL of cabbage juice in a beaker. Dropwise, with stirring, add 0.1 M NaOH until a lime-green color is obtained.

▲   Hydrochloric acid, 1.0 M HCl: Slowly adding acid to water, mix 8.3 mL of concentrated acid with enough water to make 100 mL of solution.

▲   Sodium hydroxide, 1.0 M NaOH: Dissolve 4.0 g in enough water to make 100 mL of solution.

▲   Calcium carbonate, powder.

2.   Half-fill a test tube with basic cabbage juice and another test tube with purple cabbage juice. Clamp the two tubes beside each other, placing the rubber tubing attached to the glass bend into the test tube with the cabbage juice. The purple cabbage juice serves as a reference.

3.   Using a long-stem pipet, deliver a pipetful of hydrochloric acid into the piece of glass tubing, maintaining pressure on the pipet bulb to force the acid onto the $CaCO_3$ and the gas out of the flask and through the other tube into the basic cabbage juice. Note any color changes.

4.   Continue to add pipetfuls of HCl, one at a time, noting the changes in the color of the cabbage juice. The juice will reach a light shade of lavender.

Concepts

▲   The purple color that results from boiling red cabbage in water is a mixture of anthocyanins, which are found in the sap of the plant. The anthocyanins are red in acid conditions, violet when neutral, and green in basic conditions.

▲   Anthocyan pigments are present in the form of glucosides. By hydrolysis, anthocyanins are converted into glucose or other monosaccharides and colored anthocyanidins. The three parent anthocyanidins are named pelargonidin, cyanidin, and delphinidin.

▲ The number of hydroxyl groups in the anthocyanidin part of the molecule, the other pigments present, and the acidity of the cell sap determine the various shades of natural colors.

▲ When hydrochloric acid and calcium carbonate react, hydrogen carbonate, HCO_3^-, ions are formed. As the concentration of these ions increases, they combine with hydrogen ions from hydrochloric acid and form unstable carbonic acid. When the solution becomes saturated with carbonic acid, carbon dioxide gas is evolved.

▲ The green basic color of cabbage juice goes through a series of color changes to a light pinkish purple. Because the carbonic acid is unstable, the concentration of acid does not reach a very high level and the red acidic color of cabbage juice is not reached.

Reactions

$$H^+(aq) + CO_3^{2-}(aq) \leftrightarrows HCO_3^-(aq)$$

$$H^+(aq) + HCO_3^-(aq) \leftrightarrows H_2CO_3(aq)$$

$$H_2CO_3(aq) \leftrightarrows H_2O(l) + CO_2(g)$$

Notes

1. Do not use a smaller Erlenmeyer flask or have the glass tubes too close to the calcium carbonate and acid. The calcium carbonate and acid might bubble up through the glass bend tube and into the cabbage juice solution.

2. To observe the effects of an acid on an indicator, add enough bromocresol purple indicator to distilled water to obtain a red color. Bubble CO_2 gas into the red solution to change the indicator to its yellow color in acid.

3. CO_2 gas can also be generated into neutral purple cabbage juice to change the color to slightly pinkish purple. Compare this color to that of the reference tube.

4. Carbonated water could be used instead of generating CO_2 gas, but illustrating to students how a gas can be generated is important.

5. Consult Investigation 59 for structures of the three anthocyanidins and the reaction of cyanin with acid and base and its hydrolysis to form cyanidin chloride and glucose.

References

Fuson, R. C.; Connor, R.; Price, C. C.; Snyder, H. R. *A Brief Course in Organic Chemistry;* Wiley: London, 1941; pp 167–168.

Kemp, D. S.; Vellaccio, F. *Organic Chemistry;* Worth: New York, 1980; pp 983–984.

Kingzett, C. T. *Chemical Encyclopedia;* Van Nostrand: New York, 1928; pp 565–567.

Lowy, A.; Harrow, B.; Apfelbaum, P. M. *Introduction to Organic Chemistry;* Wiley: New York, 1945; p 385.

Nebergall, W. H.; Schmidt, F. C.; Holtzclaw, H. F., Jr. *College Chemistry*, 5th ed.; Heath: Lexington, MA, 1976; p 448.

Wertheim, E. *Textbook of Organic Chemistry;* Blakiston: Philadelphia, PA, 1939; pp 650–652.

Changing the Color of Flowers with Ammonia Vapor

When calcium oxide and ammonium chloride are heated, the ammonia vapor produced changes the color of flower petals.

Procedure

Please consult the Safety Information before proceeding.

1. Place about one-fourth of the solid mixture in a Pyrex test tube.
2. Using a Bunsen burner, heat the solids in the test tube, holding a colored flower at the opening of the test tube.
3. The flower petals exposed to the vapor will change color.

Safety Information

1. Calcium oxide is corrosive through inhalation, skin and eye contact, and ingestion.
2. Ammonium chloride is toxic by ingestion.

Materials

▲　Solid: Mix together 20 g of calcium oxide and 20 g of ammonium chloride. In a mortar and pestle, grind them together into a fine powder, if necessary.

▲　Flowers (see Notes).

Concepts

▲　Plant pigments occur in plastids (specialized organelles) or in solution in the sap of the plant. Chlorophyll and carotene are found in the plastids.

▲　Two classes of sap pigments are anthoxanthins and anthocyanins. Anthoxanthins are bright yellow in alkaline solution. Anthocyanins can vary from dull red or reddish brown to purple and green.

▲　Plants can contain both anthoxanthins and anthocyanins or a mixture of just anthocyanins. Changes in the acidity of the cell sap can cause various shades of color.

▲　Exposing the flowers to aqueous ammonia vapor results in their color in basic or alkaline solution.

▲　Flowers we used and their colors before and after were as follows:

Flower	Color Before	Color After
Carnation	Red	Black
Carnation	Pink	Yellow
Chrysanthemum	Yellow	Orange
Chrysanthemum	Purple	Green

Reactions

$$CaO(s) + 2NH_4Cl(s) \xrightarrow{\Delta} 2NH_3(g) + CaCl_2(s) + H_2O(l)$$

Notes

1.　Most florists have flowers they are going to discard and usually will give them to you, especially if you tell them they will be used for science class.

2.　Try different flowers.

References

The Chemcraft Book; Porter Chemical: Hagerstown, MD, 1928; p 27.

Kingzett, C. T. *Chemical Encyclopedia;* Van Nostrand: New York, 1928; pp 565–567.

Lippy, J. D., Jr.; Palder, E. L. *Modern Chemical Magic;* Stackpole: Harrisburg, PA, 1959; p 104.

Energy Changes

Synthesis of Sodium Peroxide Octahydrate

Hydrogen peroxide is slowly added to a solution of sodium hydroxide. An exothermic reaction occurs as white crystals of sodium peroxide octahydrate form.

Procedure

Please consult the Safety Information before proceeding.

1. Place 60 mL of sodium hydroxide solution in a 125-mL flask. Place a thermometer in the solution, noting the temperature.

2. Very slowly add 40 mL of hydrogen peroxide solution down the inside of the flask. Observe before mixing the solutions together. You should see crystals at the interface between the two solutions.

3. Mix the solutions together by swirling. Note the warmth of the flask where the crystals have formed, as well as the temperature.

Safety Information

1. Sodium hydroxide is corrosive to all tissues. Inhalation of the dust or concentrated mist may cause damage to the respiratory tract. Wear gloves, goggles, and a rubber apron when handling.

2. Hydrogen peroxide, 30%, is a strong oxidant and should be used with caution. Avoid contact with the skin and eyes and wear rubber gloves and goggles.

Materials

▲ Sodium hydroxide, NaOH: Dissolve 40 g in 80 mL of water. Let cool to room temperature before using.

▲ Hydrogen peroxide, 12%, H_2O_2: Dilute 40 mL of 30% H_2O_2 with 60 mL of water.

4. Leave the flask undisturbed for further observations.

Concepts

▲ The formation of sodium peroxide octahydrate is exothermic. More energy is released than absorbed, so the change in energy for the reaction, or ΔH value, is negative.

▲ ΔH can be calculated by using the following ΔH_f° values:

Substance	$\Delta H_f^\circ(kJ/mol)$
$Na^+(aq)$	–240.12
$OH^-(aq)$	–229.994
$H_2O_2(aq)$	–191.17
$H_2O(l)$	–285.83
$Na_2O_2 \cdot 8H_2O(aq)$	–2943.0

ΔH_{rxn}° for the following reaction can be calculated as follows:

$\Delta H_f^\circ Na_2O_2 \cdot 8H_2O(s) - [2H_f^\circ Na^+(aq) +$

$\qquad 2H_f^\circ OH^-(aq) + 6\Delta H_f^\circ H_2O(l) + \Delta H_f^\circ H_2O_2(aq)]$

$= -96.6$ kJ/mol $Na_2O_2 \cdot 8H_2O(s)$

▲ The sodium peroxide octahydrate crystals belong to the monoclinic system. The unit cell is body centered, and it contains four molecules.

▲ On exposure to air, the crystals rapidly lose their brilliancy due to the absorption of carbon dioxide and evolution of oxygen. Sodium carbonate is the solid being formed.

▲ In 1967 a patent was obtained by Dow Chemical Company because James Leddy and Dale Schechter developed a method of stabilizing the crystals. They found that the crystals decompose primarily due to the presence of surface water. They found that several solvents such as diethyl ether, dioxane, tetrahydrofuran, acetonitrile, ethyl bromide, isopropyl ether, and chloroform could be used to wash the crystals but would be inert with respect to the reaction and would not remove water of hydration.

Reactions

$$2Na^+(aq) + 2OH^-(aq) + 6H_2O(l) + H_2O_2(aq) \rightarrow$$
$$Na_2O_2 \cdot 8H_2O(s) + 96.6\ kJ$$

Notes

1. The solid could be filtered and then shown to the students.

2. If you plan to store the sodium peroxide octahydrate, it should be kept in an airtight container.

References

Balej, J.; Spalek, O. *Collect. Czech. Commun.* **1982,** *47,* 736.

Fairley, T. *J. Chem. Soc.* **1877,** *31,* 125.

Harcourt, V. *J. Chem. Soc.* **1862,** *14,* 278.

J. Phys. Chem. Ref. Data **1982,** *11,* suppl. 2.

Leddy, J. J.; Schechter, D. L. U.S. Patent 3,304,618, 1967.

Mellor, J. W. *A Comprehensive Treatise on Inorganic and Theoretical Chemistry;* Wiley: New York, 1961; Vol. II, pp 632–633.

Newth, G. S. *Chemical Lecture Experiments;* Longmans, Green, & Co.: New York, 1892; p 76.

Penneman, R. A. *Inorg. Syn.* **1950,** *3,* 1.

Freezing Sodium Sulfate Decahydrate and Hydrochloric Acid

Adding sodium sulfate decahydrate crystals to a cooled solution of hydrochloric acid in a beaker and then placing the beaker in a pool of water on a wooden block causes the beaker to freeze to the block.

Procedure

Please consult the Safety Information before proceeding.

1. In a 50-mL cylinder, chill 30 mL of concentrated hydrochloric acid to 10 °C in an ice bath. Chilling will happen rather quickly.

2. Pour the acid on 60 g of sodium sulfate decahydrate crystals in a 250-mL beaker and stir to mix thoroughly.

3. Place the beaker on a wooden block covered with a pool of water. The beaker will freeze to the block in about a minute.

Safety Information

1. Sodium sulfate decahydrate may be harmful by inhalation, ingestion, or skin absorption. It causes skin irritation.

2. For hydrochloric acid, work in a fume hood and wear goggles, gloves, and a rubber apron. External contact causes severe burns, and contact with the eyes may result in a loss of vision. Inhalation causes coughing and choking with possible inflammation and ulceration of the respiratory tract.

Materials

▲ Sodium sulfate decahydrate, $Na_2SO_4 \cdot 10H_2O$, crystals: Prepare a saturated solution by slightly warming 500 mL of water on a magnetic stirrer. Add sodium sulfate decahydrate until no more will dissolve and then add a little more. Let this mixture sit overnight. There should be a good crop of crystals the next day, most of which can be removed, blotted dry, and placed in a sealed bottle until ready to use. Continue to let the water evaporate from the beaker and collect more crystals. Just before using, place the crystals between two paper towels and pulverize them with a hammer. Store any unused crystals in a bottle.

▲ Concentrated hydrochloric acid.

Concepts

▲ This is an endothermic reaction that overall requires more energy than is released. The enthalpy, or ΔH value, is positive. The energy term is included on the reactant side of the equation.

▲ Endothermic reactions undergo a decrease in temperature, and the reaction vessel feels cold to the touch.

▲ The enthalpy value is +79.45 kJ as calculated from the following individual enthalpy values (kilojoules per mole): sodium sulfate decahydrate, −4327.26; hydrochloric acid, −167.16; Na^+, −240.12; SO_4^{2-}, −909.27; Cl^-, −167.16; and H_2O, −285.83. ΔH_{rxn}° can be calculated as follows:

$$2\Delta H_f^{\circ}\, Na^+(aq) + \Delta H_f^{\circ}\, SO_4^{2-}(aq) + \Delta H_f^{\circ}\, H^+(aq) +$$
$$\Delta H_f^{\circ}\, Cl^-(aq) + 10\Delta H_f^{\circ}\, H_2O(l) -$$
$$[\Delta H_f^{\circ}\, Na_2SO_4 \cdot 10H_2O(s) + \Delta H_f^{\circ}\, HCl(aq)]$$

▲ What drives this reaction is an increase in entropy, or the number of ways in which particles can be arranged. A solid plus a solution is producing 15 species.

▲ Sodium sulfate decahydrate is also known as Glauber's salt. Glauber was an alchemist who used it as medicine in the 17th century.

▲ Crystallizing sodium sulfate at temperatures below 32.28 °C produces the decahydrate. Above 32.28 °C, the anhydrous salt, Na_2SO_4, crystallizes.

Reactions

$$79.45\ kJ + Na_2SO_4 \cdot 10H_2O(s) + HCl(aq) \rightarrow$$
$$2Na^+(aq) + SO_4^{2-}(aq) + H^+(aq) +$$
$$Cl^-(aq) + 10H_2O(l)$$

Notes

1. When the amounts indicated in the demonstra-

tion are used, the temperature will drop to about –5 °C whether or not the HCl is cooled, but the beaker will not freeze to the block of wood unless the HCl is first cooled to 10 °C.

2. Beginning the demonstration by cooling about 5 mL of concentrated acid to 10 °C and carefully adding this solution to about 10 mL of water will be a good contrast for what happens in the demonstration. The temperature increase will be about 10 °C. Be sure to record the initial and final temperatures of the water.

3. Sodium sulfate crystals cannot sit in the air too long or they will begin to effloresce, becoming white and brittle.

4. Another demonstration is to fill a 250-mL beaker with a salt–ice–water mixture and freeze it to the block of wood. Use about 115 g of ice, 50 mL of water, and about 30 g of table salt (NaCl). Stir to obtain a good slush and record the temperature. The temperature drops well below –10 °C. The two demonstrations could be compared (both endothermic) and contrasted (chemical versus physical reaction).

References

Benedict, F. G. *Chemical Lecture Experiments;* Macmillan: London, 1901; pp 352–353.

CRC Handbook of Chemistry and Physics, 72nd ed.; Lide, D. R., Ed.; CRC: Boca Raton, FL, 1992; pp 5-17–5-59.

Nebergall, W. H.; Schmidt, F. C.; Holtzclaw, H. F., Jr. *College Chemistry*, 5th ed.; Heath: Lexington, MA, 1976; p 947.

Temperature-Sensitive Iodomercurate Complexes

Silver and copper(I) iodomercurates are synthesized. When enclosed in sealed glass tubes and heated, they change color. They revert to their original colors when cooled.

Procedure

Please consult the Safety Information before proceeding.

Preparation of Mercury Tubes

1. Using a 250-mL beaker, slowly in increments of about 5 mL each, mix 50 mL of potassium iodide solution with 50 mL of mercury(II) chloride solution.

2. Using two separate beakers, divide the solution into two volumes of 50 mL each.

3. To one beaker, add 11.4 g of copper(II) sulfate pentahydrate, and to the second, add 8.6 g of silver nitrate. Stir the solutions to mix thoroughly.

Materials

▲   Mercury(II) chloride, 0.5 M $HgCl_2$: Dissolve 3.4 g in enough water to make 50 mL of solution.

▲   Potassium iodide, 0.5 M KI. Dissolve 4.2 g in enough water to make 50 mL of solution.

▲   Copper(II) sulfate pentahydrate, powder.

▲   Silver nitrate, powder.

4.   Filter each of the precipitates and dry overnight in a desiccator.

5.   Cut two pieces of Pyrex glass tubing, 30 cm long with a 5–6-mm inner diameter. Wearing thermal gloves, seal one end by heating the tube to red hot, pinching the tubing closed with tweezers. **CAREFUL! HOT!** Let the tubes cool thoroughly.

6.   Wearing gloves and using a microspatula, transfer each precipitate into a glass tube. Seal the other end of each tube closed by heating to red hot and pinching closed with tweezers.

Procedure for Demonstration

1.   Wearing thermal gloves, or holding each glass tube with tweezers, *gently* heat the tube in a burner flame and observe the color changes. Heat just enough to have some of the compound change color.

2.   Let each tube cool to room temperature to return to its original color.

Concepts

▲   During the mixing of the $HgCl_2$ and KI solutions, a fluorescent orange precipitate of HgI_2 forms. The mercury(II) iodide solid redissolves and forms mercury(II) iodide in solution as additional KI(aq) is added.

▲   Mercury(II) iodide plus additional iodide ions form the complex ion HgI_4^{2-}, called tetraiodomercurate(II).

▲   Following are the colors of the precipitates before and during heating:

Name	Compound	Color Before	Color During
Silver tetraiodomercurate(II)	Ag_2HgI_4	Yellow	Orange–brown
Copper(I) tetraiodomercurate(II)	Cu_2HgI_4	Red	Black–grey

▲   The general reaction that is occurring is a double decomposition reaction:

$$MA + K_2HgI_4 \leftrightharpoons KA + MIm$$

where MA is a soluble metallic salt and Im designates one of the iodomercurate complexes.

▲ One of the first references to the synthesis of an iodomercurate is that of Boullay. He combined a hot aqueous solution of ammonium iodide with mercury(II) iodide, forming hydrated ammonium triiodomercurate(II).

▲ Three- and four-coordinate complexes are formed with the large iodine atom and mercury(II). Silver and copper(I) form Hg(II) iodides, which exist in α and β forms, high and low temperature, respectively. Ag_2HgI_4 changes above 50.7 °C and Cu_2HgI_4 above 70 °C.

▲ The structures of $\beta\text{-}Cu_2HgI_4$ and $\beta\text{-}Ag_2HgI_4$ are as follows:

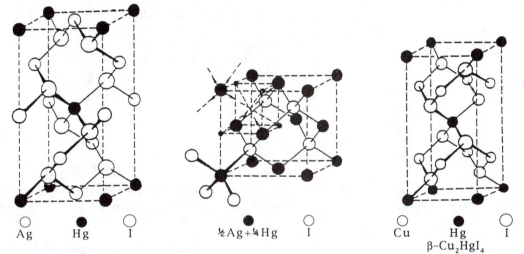

Reproduced with permission from A. F. Wells, Structural Inorganic Chemistry, *3rd ed.*; Oxford University Press, 1962.

▲ In the β form of silver tetraiodomercurate(II), the iodine ligand atoms are closer to the Hg than the Ag. Heating causes the iodine ligands to come closer to the Ag, and the solid becomes orange.[1] The reaction is similar for copper(I) tetraiodomercurate(II).

▲ The two compounds used in this demonstration are examples of thermochromic compounds— ones that change color with temperature. Both compounds are used in temperature-indicating paints. The most established use for thermochromic materials is the irreversible system that

---

[1]Nelson, R., Trenton State College, personal communication, June 1991.

leaves a record of heat distribution on an engine part. About 2000 thermochromic compounds have been identified for use in chemical analysis.

▲   A patent was issued in 1966 for a group of organic compounds, quinones with an $-NHCH_2CH_2OH$ substituent, that are light in color when cool but deeply colored when heated. These compounds are used in the paper or ink of official documents so they can be easily tested for genuineness (German patent).

Reactions

$$Hg^{2+}(aq) + 2I^-(aq) \rightarrow HgI_2(s)$$
$$HgI_2(s) \rightarrow HgI_2(aq)$$
$$HgI_2(aq) + 2I^-(aq) \rightarrow HgI_4{}^{2-}(aq)$$
$$2Cu^+(aq) + HgI^{2-}(aq) \leftrightarrows Cu_2HgI_4(s)$$
$$2Ag^+(aq) + HgI_4{}^{2-}(aq) \leftrightarrows Ag_2HgI_4(s)$$

Notes

1.   Doing Step 1 under Preparation of Mercury Tubes slowly will allow the formation of $HgI_2(s)$ to be observed.

2.   The copper and silver iodomercurates return to their original colors fairly quickly.

3.   Other iodomercurates could be made by using bismuth, cadmium, lead, or antimony salts. Consult Meyer (1943).

4.   Once the tubes are prepared, they can be used over and over again.

References

Boullay, P. F. G. *Ann. Chim. Phys.* **1827**, *34*, 345.

Comprehensive Inorganic Chemistry; Pergamon, New York, 1973; Vol. 3, pp 302–303.

Day, J. In *Kirk-Othmer Encyclopedia of Chemical Technology,* 3rd ed.; Wiley: New York, 1979; Vol. 6, pp 130–134.

German Patent 1,228,972, 1966.

Hughes, W. J. *J. Chem. Educ.* **1965**, *42*, A413.

Mellor, J. W. *A Comprehensive Treatise on Inorganic and Theoretical Chemistry;* Longmans, Green, & Co.: London, 1923; Vol. 4, pp 925–942 (includes an extensive bibliography dating from 1827: silver, pp 932, 937–938; copper, p 935).

Meyer, M. *J. Chem. Educ.* **1943**, *20*, 145–146.

Nebergall, W. H.; Schmidt, F. C.; Holtzclaw, H. F., Jr. *College Chemistry*, 5th ed.; Heath: Lexington, MA, 1976; p 848.

Singer, W.; Nowack, M. In *Kirk-Othmer Encyclopedia of Chemical Technology,* 3rd ed.; Wiley: New York, 1981; Vol. 15, pp 160–161.

Wells, A. F. *Structural Inorganic Chemistry*, 4th ed.; Clarendon: Oxford, England, 1975; p 632.

A Cold Mixture of Sodium Sulfate Decahydrate and Ammonium Nitrate

Crystals of sodium sulfate decahydrate and ammonium nitrate powder are mixed in a beaker. The result is a syrupy-looking mixture whose temperature falls to just below freezing.

Procedure

Please consult the Safety Information before proceeding.

1. In a beaker, mix about 40 g of sodium sulfate decahydrate and 40 g of ammonium nitrate solids until syrupy.

2. Record the temperature after 2 min. It will be about –2 °C.

Concepts

▲ The reaction is endothermic and requires more energy than is released. Its enthalpy, or ΔH value, is positive.

Safety Information

1. Ammonium nitrate is an irritant through inhalation and skin and eye contact. It is moderately toxic by ingestion.

2. Sodium sulfate decahydrate may be harmful by inhalation, ingestion, or skin absorption. It causes skin irritation.

Materials

▲ Sodium sulfate deca-hydrate, $Na_2SO_4 \cdot 10H_2O$: Prepare a saturated solution by slightly warming and stirring 500 mL of water on a hot plate–stirrer. Add sodium sulfate decahydrate until no more solid will dissolve and then add a little more. Let this solution sit overnight. There should be a good crop of crystals the next day, most of which can be removed, blotted dry, and placed in a sealed bottle until ready to use. Continue to let the water evaporate from the beaker and collect more crystals. Just before using, place the crystals between two paper towels and pulverize with a hammer. Using a hammer is easier than using a mortar and pestle. Store any unused crystals in a bottle.

▲ Ammonium nitrate, powder.

▲ The enthalpy value is +578.31 kJ, as calculated from the following individual enthalpy values (kilojoules per mole): sodium sulfate decahydrate, –4327.26; sodium nitrate, –447.49; ammonium nitrate, –365.56; ammonium sulfate, –1174.28; and water, –285.83. ΔH°_{rxn} can be calculated as follows:

$$2\Delta H^\circ_f NaNO_3(aq) + \Delta H^\circ_f (NH_4)_2SO_4(aq) +$$
$$10\Delta H^\circ_f H_2O(l) - [\Delta H^\circ_f Na_2SO_4 \cdot 10H_2O(s) +$$
$$2\Delta H^\circ_f NH_4NO_3(s)]$$

▲ Endothermic reactions exhibit a decrease in temperature when they occur, and the reaction vessel feels cold to the touch. The temperature of this reaction reaches –2 to –3 °C.

▲ Entropy, or ΔS, is a measure of the amount of randomness of a system or of the ways in which the particles can be arranged. The process of the two solids forming an aqueous mixture results in an increase in the entropy of the system. The reaction is favored because of an increase in entropy.

▲ Crystallizing sodium sulfate at temperatures below 32.28 °C produces the decahydrate. Above the transition of 32.28 °C, the anhydrous salt, Na_2SO_4, crystallizes.

▲ Sodium sulfate decahydrate is also known as Glauber's salt. Glauber was an alchemist who used it as medicine in the 17th century.

Reactions

$$\text{energy} + Na_2SO_4 \cdot 10H_2O(s) + 2NH_4NO_3(s) \rightarrow$$
$$2NaNO_3(aq) + (NH_4)_2SO_4(aq) + 10H_2O(l)$$

Notes

Be sure to pulverize the sodium sulfate crystals before using. A smaller crystal size ensures a larger drop in temperature.

References

CRC Handbook of Chemistry and Physics, 72nd ed.; Lide, D. R., Ed.; CRC: Boca Raton, FL, 1992; pp 5-17–5-59.

Lippy, J. D., Jr.; Palder, E. L. *Modern Chemical Magic;* Stackpole Co.: Harrisburg, PA, 1959; p 14.

Natl. Bur. Stand. (U.S.) **1952**, *Feb 1.*

Nebergall, W. H.; Schmidt, F. C.; Holtzclaw, H. F., Jr.; *College Chemistry*, 5th ed.; Heath: Lexington, MA, 1976; p 947.

Luminol and Dimethyl Sulfoxide

White pellets of potassium hydroxide are placed in a colorless liquid containing luminol; the pellets begin to glow in a darkened room. When the flask is shaken, the liquid begins to glow.

Procedure

Please consult the Safety Information before proceeding.

1. Place 20 g of potassium hydroxide pellets in a 400-mL Erlenmyer flask.

2. Add 150 mL of luminol–dimethyl sulfoxide solution and stopper.

3. Turn off the lights and observe the glowing pellets.

4. Carefully shake the stoppered flask, and the solution will begin to glow.

Safety Information

1. Dimethyl sulfoxide is readily absorbed through the skin and results in primary irritation with redness, itching, and sometimes scaling. It can transport materials through the skin. Contact with the skin or eyes must be avoided; wear rubber gloves and eye protection and wash your hands thoroughly after use.

2. Toxic and carcinogenic properties of luminol, 3-aminophthalic hydrazide, are not known.

3. Potassium hydroxide is corrosive to all tissues. Inhalation of the dust or concentrated mist may cause damage to the respiratory tract.

Materials

▲ Potassium hydroxide pellets, KOH.

▲ Luminol–dimethyl sulfoxide, $C_6H_3NH_2(CO)_2(NH)_2$–$(CH_3)_2SO$: Add 0.5 g of luminol to 150 mL of dimethyl sulfoxide.

5. Flush the flask with pure oxygen for a brighter and longer glow.

6. After several days the flask will glow even brighter when shaken, the glow lasting for 10–15 min.

Concepts

▲ Dimethyl sulfoxide is an aprotic solvent, that is, a polar solvent of moderately high dielectric constant. The larger the magnitude of the dielectric constant, the greater the polar character of the solvent. Aprotic solvent reactions tend to proceed rapidly to produce high yields of products at room temperature. Thus, it is an excellent solvent for this system.

luminol
(3-aminophthalic hydrazine)

dianion of luminol

▲ In basic solution, the bonds between the nitrogen and hydrogen atoms of the two –NH groups on the luminol molecule are broken, producing the dianion of luminol. The hydrogen combines with the OH⁻ ions to form water.

▲ Dimethyl sulfoxide causes the luminol dianion to form the excited-state product, and N_2 gas is released. The excited-state product decays to the ground state, emitting energy in the form of light.

▲ Molecular oxygen dissolved in the dimethyl sulfoxide converts the dianion to the excited state.

Reactions

| dimethyl sulfoxide | excited state of 3-aminophthalic ion | ground state of 3-aminophthalic ion |

Notes

1. The luminol is best added just after the dimethyl sulfoxide is added to the flask.

2. After several days, the dimethyl sulfoxide will gradually darken. This color change appears not to affect the brightness. As the solution sits for several days, the glow appears to brighten and becomes bluer.

3. Aerate by opening, closing, and then shaking the flask.

References

Chalmers, H. H.; Bradbury, M. C.; Fabricant, J. D. *J. Chem. Educ.* **1987**, *64*, 969.

Ealy, Julie B.; Ealy, James L., Jr. *Close-up on Chemistry;* ACS video manuscript; American Chemical Society: Washington, DC, 1991; pp 19–20.

Morrison, R. T.; Boyd, R. N. *Organic Chemistry*, 3rd ed.; Allyn and Bacon: Boston, MA, 1973; pp 31–32.

Rosewell, D. F.; White, E. H. *Meth. Enzym.* **1978**, *57*, 409.

Schneider, H. W. *J. Chem. Educ.* **1941**, *18*, 347.

Schneider, H. W. *J. Chem. Educ.* **1970**, *47*, 519.

White, E. H. *J. Chem. Educ.* **1957**, *34*, 275.

Solutions and Solubility

Silver Compounds

To a colorless solution of silver nitrate, other solutions are added. Various colored compounds appear and disappear.

Procedure

Please consult the Safety Information before proceeding.

1. Place 100 mL of silver nitrate solution in a 250-mL beaker.
2. Add 2 drops of sodium carbonate solution. A cream precipitate swirls to the bottom.
3. Add 1 drop of sodium hydroxide solution. A brown precipitate swirls to the bottom and mixes with the cream precipitate.
4. Add 1 drop of sodium chloride solution. A white precipitate swirls to the bottom.

4. Inhalation of concentrated ammonia causes edema of the respiratory tract, spasm of the glottis, and asphyxia. Treatment must be prompt to prevent death.

5. Ammonium sulfide is a poison by skin contact and subcutaneous, intravenous, intradermal, parenteral, and intraperitoneal routes. It is also pyrophoric in air.

6. Sodium carbonate, potassium thiocyanate, and sodium iodide may be harmful by inhalation, ingestion, or skin absorption. They cause eye and skin irritation. They are irritating to mucous membranes and the upper respiratory tract.

7. Sodium bromide may be harmful by inhalation, ingestion, or skin absorption. Exposure can cause sedative, hypnotic, and anticonvulsant effects.

5. Dropwise, with stirring add about 15 drops of concentrated ammonia until clearing occurs.

6. Add 1 drop of sodium bromide solution. A white precipitate forms.

7. With stirring, add 8 mL of concentrated ammonia until clearing occurs.

8. Add 1 drop of sodium iodide solution. A yellow–green precipitate forms and swirls to the bottom.

9. Add 1 drop of potassium thiocyanate solution. A dense white precipitate forms.

10. With stirring, add 16 mL of concentrated ammonia. This mixture will become relatively clear, but some yellow specks of precipitate will remain.

11. Add 1 drop of ammonium sulfide solution. A thick black precipitate forms.

Concepts

▲ For compounds slightly soluble in water, the equilibrium constant is called a solubility product constant, K_{sp}. For a saturated solution of the compound, the quantity is constant at constant temperature.

▲ The following are K_{sp} (solubility product constant) values:

Compound	K_{sp}
Ag_2CO_3	8.45×10^{-12}
$AgCl$	1.77×10^{-10}
$AgBr$	5.35×10^{-13}
AgI	8.51×10^{-17}
$AgSCN$	1.03×10^{-12}
Ag_2S	1.09×10^{-49}

In order for precipitation to take place, the product of the ion concentrations (P) must exceed the solubility product. For example, K_{sp} for Ag_2CO_3 is 8.45×10^{-12}

$$Ag_2CO_3(s) \rightarrow 2Ag^+(aq) + CO_3^{2-}(aq)$$

Calculating P, the ion product:

$$[CO_3^{2-}] = \frac{1.0 \text{ mol}}{1 \text{ L}} \times \frac{1}{100 \text{ mL}} \times 2 \text{ drops} \times$$

$$\frac{1 \text{ mL}}{20 \text{ drops}} = 0.001 \text{ M}$$

$[Ag^+] = 0.002 \text{ M}$

$P = [Ag^+]^2[CO_3^{2-}] =$

$$[0.002]^2[0.001] = 4.0 \times 10^{-9}$$

Because the product of the ion concentrations, 4.0×10^{-9}, exceeds the solubility product, 8.45×10^{-12}, a precipitate of sodium carbonate will form.

▲ The equilibrium constant for each of the individual reactions can be calculated by using the K_{sp} constants. For an explanation of some of the reactions, see Shakhashiri (1983).

▲ Ammonia is often used as a complexing agent. Silver ions, Ag^+, are converted to a complex ion such as $Ag(NH_3)_2^+$. Complexing the silver ions prevents them from being able to precipitate with the carbonate, hydroxide, chloride, bromide, iodide, or thiocyanate ions.

▲ The dissociation constant for $Ag(NH_3)_2^+$ is 4×10^{-8}. This value indicates that the products, Ag^+ and NH_3, are not favored at equilibrium but the complex is.

Materials

▲ Silver nitrate, 0.02 M $AgNO_3$: Dissolve 1.7 g in enough distilled water to make 500 mL of solution. Store in a brown bottle to prevent darkening.

▲ Ammonium sulfide, 23.9% NH_4S: A yellow commercial stock solution.

▲ The following solutions are all 1.0 M. Dissolve the specified amount in enough distilled water to make 100 mL of solution:

— Sodium carbonate, Na_2CO_3: 10.6 g.

— Sodium hydroxide, NaOH: 4.0 g.

— Sodium chloride, NaCl: 5.8 g.

— Sodium bromide, NaBr: 10.3 g.

— Sodium iodide, NaI: 14.3 g.

— Potassium thiocyanate, KSCN: 9.7 g.

Reactions

The numbers correspond to the steps under Procedure.

2. $2Ag^+(aq) + CO_3^{2-}(aq) \rightarrow Ag_2CO_3(s)$

3. $Ag_2CO_3(s) + 2OH^-(aq) \rightarrow 2AgOH(s) + CO_3^{2-}(aq)$

4. $AgOH(s) + Cl^-(aq) \rightarrow AgCl(s) + OH^-(aq)$

5. $AgCl(s) + 2NH_3(aq) \rightarrow Ag(NH_3)_2^+(aq) + Cl^-(aq)$

6. $Ag(NH_3)_2^+(aq) + Br^-(aq) \rightarrow AgBr(s) + 2NH_3(aq)$

7. $AgBr(s) + 2NH_3(aq) \rightarrow Ag(NH_3)_2^+(aq) + Br^-(aq)$

8. $Ag(NH_3)_2^+(aq) + I^-(aq) \rightarrow AgI(s) + 2NH_3(aq)$

9. $AgI(s) + SCN^-(aq) \rightarrow AgSCN(s) + I^-(aq)$

10. $AgSCN(s) + 2NH_3(aq) \rightarrow Ag(NH_3)_2^+(aq) + SCN^-(aq)$

11. $2Ag(NH_3)_2^+(aq) + S^{2-}(aq) \rightarrow Ag_2S(s) + 4NH_3(aq)$

Notes

1. The addition of ammonia can be done without a magnetic stirrer, although it is more convenient to use one.

2. The demonstration can be performed on an overhead projector, but simply use the light of the projector to better display the various precipitates formed. A more diffuse light source is even better.

3. Storing the solutions in small dropper bottles makes them readily available and easy to use whenever you want to perform the demonstration.

4. Place 10 mL of silver nitrate solution in each of seven test tubes. Add 5 drops of each solution, except ammonia, to produce the seven precipitates so they can be separately displayed.

Reference

CRC Handbook of Chemistry and Physics, 72nd ed.; Lide, D. R., Ed.; CRC: Boca Raton, FL, 1992; p 8-43.

Masterton, W. L.; Slowinski, E. J.; Stanitski, C. L. C*hemical Principles*, alternate ed.; CBS College: 1983; pp 602–603.

Schwenck, J. R. *J. Chem. Educ.* **1959**, *36*, 45.

Shakhashiri, B. Z. *Chemical Demonstrations; A Handbook for Teachers of Chemistry*, 1st ed.; University of Wisconsin Press: Madison, WI, 1983; Vol. I, pp 307–313.

Visible Conservation of Mass

Solutions of bismuth nitrate and potassium bromide are used in this demonstration. One solution is placed in a stoppered test tube, and one is left free in a small plastic bag. Upon mixing the two solutions, a chemical reaction takes place and a light yellow solid forms.

Procedure

Please consult the Safety Information before proceeding.

1. Place 8 mL of one solution in a small reclosable plastic bag. Close and weigh the bag.

2. Place 8 mL of the other solution in a small test tube. Stopper and weigh the test tube.

3. Place the stoppered test tube in the plastic bag and close the bag.

4. Unstopper the test tube and mix the two solutions.

5. Weigh the plastic bag.

Safety Information

1. Bismuth nitrate and bismuth bromide are moderately toxic by the intraperitoneal route.

2. If large doses of potassium bromide are taken internally, central nervous system depression will result. Prolonged intake may cause mental deterioration and acneform skin eruptions.

3. Continued exposure to the vapor of nitric acid may cause chronic bronchitis. Use a fume hood, and wear gloves, goggles, and a protective apron when handling.

Materials

▲   Bismuth nitrate, $Bi(NO_3)_3$: Dissolve 3.95 g in 100 mL of water, acidified with 4 drops of concentrated nitric acid.

▲   Potassium bromide, KBr: Dissolve 3.57 g in 100 mL of water.

Concepts

▲   Mass is conserved in a chemical reaction. The mass of the starting substances, the reactants, is equal to the mass of the final substances, the products.

▲   Antoine Lavoisier is usually credited with the Law of Conservation of Mass. In 1789 he formally stated the "law of indestructibility of matter", but quantitative methods in chemistry were used before his time by van Helmont, Boyle, and Black. In fact, Lavoisier used the researches of Black as his model. Johann Baptista van Helmont, who published in 1648, was quantitative in his work, made extensive use of the balance, and clearly expressed the law of indestructibility of matter.

▲   Traditionally, lead nitrate and potassium iodide are used to make the yellow solid lead iodide. Because lead nitrate is poisonous, bismuth nitrate is substituted here. The solid formed does not have the bright yellow color of lead iodide but instead is light yellow. Industrially, bismuth is considered one of the less toxic of the heavy metals.

Reactions

1. $Bi(NO_3)_3(aq) + 3KBr(aq) \rightarrow BiBr_3(s) + 3KNO_3(aq)$
2.   Net equation for the formation of the precipitate:

$$Bi^{3+}(aq) + 3Br^-(aq) \leftrightharpoons BiBr_3(s)$$

Notes

1.   Other solutions that combine to form precipitates could also be used.

2.   Although the demonstration could be done in any self-contained unit, the plastic bag works nicely, and the reaction is easily visible to all students.

3. Each student could be given a plastic bag, and several different pairs of solutions could be used throughout the classroom.

References

Fowles, G. *Lecture Experiments in Chemistry;* Blakiston: Philadelphia, PA, 1937; p 25.

Partington, J. R. *A Short History of Chemistry;* Dover: New York, 1989; pp 45–46, 152, 177.

Efflorescence

Brittle icelike crystals of sodium carbonate decahydrate, when exposed to the air, become opaque white and appear to be coated with a flaky substance.

Procedure

Please consult the Safety Information before proceeding.

1. Remove some crystals from the saturated sodium carbonate solution.
2. Place them on a paper towel and blot off the water.
3. Place the crystals on a watch glass and observe the changes that occur.

Materials

▲ Sodium carbonate decahydrate, $Na_2CO_3 \cdot 10H_2O$: Make a saturated solution by dissolving 32 g in 50 mL of H_2O. Add an additional 10 g. Let the mixture sit overnight so the crystals can grow.

Concepts

▲ Some hydrated salts, when exposed to air, lose a part of their water of crystallization. This process is called efflorescence. It changes the icelike crystals to opaque white crystals.

▲ Sodium sulfate decahydrate and sodium carbonate decahydrate both exhibit this property.

▲ Sodium carbonate decahydrate loses 9 of its 10 water molecules and forms sodium carbonate monohydrate, $Na_2CO_3 \cdot H_2O$. A change in the crystalline form occurs.

▲ The vapor pressure of the decahydrate salt is greater than atmospheric pressure, so the water evaporates from the decahydrate salt.

Reactions

$$Na_2CO_3 \cdot 10H_2O(s) \rightarrow Na_2CO_3 \cdot H_2O(s) + 9H_2O(g)$$

Notes

1. It takes about 5–7 min for the crystals to start to become opaque white.

2. To demonstrate efflorescence using sodium sulfate decahydrate, dissolve 40 g in 60 mL of water. Let this sit overnight and remove crystals the next day.

References

Kingzett, C. T. *Chemical Encyclopedia;* Van Nostrand: New York, 1928; p 229.

Nebergall, W. H.; Schmidt, F. C.; Holtzclaw, H. F., Jr. *College Chemistry*, 5th ed.; Heath: Lexington, MA, 1976; p 871.

Newth, G. S. *Chemical Lecture Experiments;* Longmans, Green, & Co.: New York, 1892; p 65.

Solution versus Solvent

The volume occupied by a solvent alone is compared with the volume of a solution.

Procedure

Please consult the Safety Information before proceeding.

1. Place 225 mL of water in a 250-mL Erlenmeyer flask. Note the level of the water with a piece of masking tape.

2. Add approximately 25 g of glucose. Stir to dissolve the glucose and note the level of the solution.

3. Using a different flask, repeat with another solid such as sodium hydrogen carbonate or copper(II) sulfate pentahydrate.

Safety Information

1. Although glucose is a food substance, no substance should ever be tasted in the laboratory.

2. Sodium hydrogen carbonate is moderately toxic by ingestion.

3. Copper(II) sulfate pentahydrate is toxic if ingested and may be harmful if it comes in contact with mucous membranes or if the dust is inhaled.

Materials

▲ Glucose, powder.

▲ Sodium hydrogen carbonate, powder.

▲ Copper(II) sulfate pentahydrate, powder.

Concepts

▲ A solution is often made up of a solid dissolved in a liquid, which acts as the solvent. The resulting solution is clear; that is, one can see through it. A clear solution can be either colorless or colored.

▲ Both the liquid and solid, separate from each other, obviously take up space; they have volume. When the solid dissolves, it essentially "disappears", but it still takes up space.

▲ An amount of 25 g of most solids that are soluble in water will show about the same height increase in volume in an Erlenmeyer flask, about 1 cm in a 250-mL flask.

Reactions

$$H_2O + solid \rightarrow solid(aq)$$

Notes

1. Showing your students how a solution is made will help them distinguish between a solution and a liquid. When students cannot "see" a solid dissolved in a liquid, they think nothing else is present except the liquid. Not understanding this point causes difficulty in a number of conceptual areas in chemistry.

2. When teaching molarity, what does "1000 mL of solution" mean? Having shown this demonstration to your students, you can make up a solution using a volumetric flask. The level of solution must not go above the etched mark on the flask. What would happen if you used 1000 mL of water and the calculated amount of solid? The solution could be above the etched mark. You would have more than 1000 mL of solution.

References

Fowles, G. *Lecture Experiments in Chemistry;* Blakiston: Philadelphia, PA, 1937; p 76.

To Precipitate or Not

The addition of hydrochloric acid to a colorless solution of sodium silicate produces a precipitate, whereas addition of sodium silicate solution to hydrochloric acid does not produce a precipitate.

Procedure

Please consult the Safety Information before proceeding.

1. Place 10 mL of sodium silicate solution in a test tube.

2. Dropwise, add 10 drops of hydrochloric acid, and observe that a white gel forms at the top.

3. Place 10 mL of hydrochloric acid in another test tube.

4. Dropwise, add 10 drops of sodium silicate solution, and observe no visible reaction.

Safety Information

1. External contact with hydrochloric acid causes severe burns, and contact with the eyes may result in permanent damage. Inhalation causes coughing and choking with possible inflammation and ulceration of the respiratory tract. Work in a fume hood, and wear goggles, gloves, and a rubber apron.

2. Sodium silicate is irritating and caustic to the skin and mucous membranes. Ingestion causes vomiting and diarrhea.

Materials

▲ Sodium silicate nonahydrate, 1.0 M $NaSiO_3 \cdot 9H_2O$: Dissolve 28.4 g in enough warm distilled water to make 100 mL of solution. Cool to room temperature before using.

▲ Hydrochloric acid, 6 M HCl: Slowly pour 50 mL of concentrated acid into 40 mL of distilled water. Cool the solution. Stir and dilute the solution to 100 mL.

Concepts

▲ When hydrochloric acid is added dropwise to the sodium silicate solution, a white gelatinous precipitate of metasilicic acid, or silica gel, forms. When sodium silicate is added to hydrochloric acid, a colloid is formed, with metasilicic acid in solution.

▲ Silica gel has an open, porous structure with a large surface area per unit of mass. It is adapted for adsorption of gases and catalysis of certain chemical reactions involving substances in the gaseous state.

▲ In a gel, the dispersed phase in a colloidal system coagulates in a manner so the whole mass, including the liquid, forms an extremely viscous gelatinous body. When a gel is formed, water or some other solvent is taken up so it is said to be hydrated or solvated. Silica gel is a colloidal dispersion of hydrated silicon dioxide or water glass.

▲ Chain metasilicates ($SiO_3{}^{2-}$) are formed by corner sharing of {SiO_4} tetrahedra. Each {SiO_4} shares two oxygen atoms with contiguous tetrahedra. In Na_2SiO_3, a repeat occurs after every second tetrahedron. The chains are stacked in parallel. This structure provides six or eight coordination sites for positive ions.

▲ Van Bemmelen (1897) observed that dried silicic acid adsorbed various vapors and gases.

Reactions

$$Na_2SiO_3 \cdot 9H_2O(aq) + 2HCl(aq) \rightarrow$$
$$2NaCl(aq) + H_2SiO_3(aq/s) + 9H_2O(l)$$

Notes

The reaction also occurs if 56 g of sodium silicate is dissolved in enough water to make 100 mL of solution. This makes a 2 M solution instead of a 1 M solution. The same amount of 6.0 M HCl would be used.

Na$_2$SiO$_3$

520 is the repeat distance in pm after 2 tetrahedra.
(Reprinted with permission from Greenwood, N.N.;
Earnshaw, A. Chemistry of the Elements, 1st ed.;
Pergamon: New York, 1984; p 404.)

References

Greenwood, N. N.; Earnshaw, A. *Chemistry of the Elements*, 1st ed.; Pergamon: New York, 1984; pp 403–404.

Nebergall, W. H.; Schmidt, F. C.; Holtzclaw, H. F., Jr. *College Chemistry*, 5th ed.; Heath: Lexington, MA, 1976; pp 324, 733.

Newth, G. S. *Chemical Lecture Experiments;* Longmans, Green, & Co.: New York, 1892; p 225.

Van Bemmelen. *Z. Anorg. Chem.* **1897**, *13*, 296.

A Gold Glyoxal Sol

Drops of a gold solution are added to a colorless solution of sodium carbonate. Upon addition of 40% glyoxal, blue–gray colloidal gold forms.

Procedure

Please consult the Safety Information before proceeding.

1. Place 20 mL of 0.5 M sodium carbonate solution in a test tube.

2. Add 10 drops of 0.1% gold solution and swirl the solution to mix.

3. Add 1 drop of 40% glyoxal and swirl the solution to mix. A blue–gray colloid slowly forms.

4. Pass the test tube around the class so that students can observe the "solution".

5. Using an overhead projector light, place a piece of paper with a small circle cut out in front of

Safety Information

1. Practice extreme care if you are using a laser in this demonstration. Avoid all direct contact with the eyes because it will cause eye damage.

2. Never use a solution of commercial glyoxal that is stronger than 40% glyoxal. Glyoxal can be absorbed through the skin and is also flammable. It is moderately irritating to the skin and mucous membranes. Work in an efficient fume hood when preparing solutions of this substance. ➔

3. Sodium carbonate causes sensitivity reactions by inhalation, ingestion, or skin absorption. It causes eye and skin irritation. It is irritating to mucous membranes and the upper respiratory tract.

4. Hydrogen tetrachloroaurate(III) trihydrate should be used only in a fume hood. It is harmful if swallowed, inhaled, or absorbed through the skin. The material is extremely destructive to tissue of mucous membranes and the upper respiratory tract, eyes, and skin.

the light to obtain a small, but direct, beam of light. Hold the test tube in front of the light. The beam of light will be scattered by the colloidal particles and appear as a faint shaft of light. Preferably, use a helium–neon laser as the light source. The scattered red beam will be seen easily. Exercise extreme care if using a laser of any kind.

Concepts

▲ Formalin, a 40% formaldehyde solution, previously was used in this demonstration instead of glyoxal. Formaldehyde is a known carcinogen, and 40% glyoxal was substituted for it. The similarity in the two substances can be seen from the structural formulas:

formaldehyde   glyoxal

▲ The term *colloid* was first used by Thomas Graham in 1861 to classify substances that usually exist in a gelatinous condition, such as starch solution and gelatin. Colloid is from the Greek words *kolla*, meaning glue, and *eidos*, meaning like.

▲ A colloidal solution consists of a fluid medium and suspended particles. The suspended particles are present in a relatively small amount. The substance present in excess constitutes the dispersion medium, and the other, the suspended particles, constitute the dispersed phase. A colloidal system consisting of a solid and liquid is called a *sol*.

▲ When the dispersed phase is of molecular size, an ordinary solution results. If larger particles are present, it is called a dispersion. A colloid lies between these two.

▲ Common examples of sols are paints and milk of magnesia. Jell-O is a gel when it is cool but a sol before it solidifies. One of the places where sols are important is in the human body. Enzymes and antibodies fold so that polar groups located on the large molecules interact with water molecules, and they are then kept in suspension.

▲ Visible light with wavelengths of 4000–7000 Å is scattered by colloidal particles of sizes ranging from 10 to 10,000 Å because of the similarity of their sizes to each other. This scattering of light is known as the Tyndall effect after John Tyndall, who first studied it in 1869. A beam of light will be visible because of scattering in a colloid but will not be visible if directed through a solution.

▲ Colloidal particles appear to be in a state of irregular rapid, dancing motion, called Brownian movement, named after Robert Brown, who first observed this motion in 1828. Colloidal particles are bombarded on all sides by the moving molecules of the dispersion medium. This irregular motion prevents the dispersed particles from settling.

Materials

▲ Sodium carbonate, 0.5 M Na_2CO_3: Dissolve 5.3 g in enough water to make 100 mL of solution.

▲ Gold solution, hydrogen tetrachloroaurate(III) trihydrate, 0.1% $HAuCl_4 \cdot 3H_2O$: Dissolve 0.1 g in enough distilled water to make 100 mL of solution. Store the solution stoppered and in the dark. Light will affect the solution.

▲ Glyoxal, 40% CHOCHO: Order as 40%. A precipitate may form in this solution because of prolonged storage but can be redissolved by warming to 50–60 °C.

▲ Preferable but not absolutely necessary: helium–neon laser light source.

Reactions

Glyoxal reduces the gold(III), and colloidal gold is formed.

Notes

1. You might want to compare a test tube containing a solution with one containing a colloid, so students can observe that light passes through a solution but not a colloid.

2. The scattered red beam from a helium–neon laser can be easily observed through Jell-O.

References

Alyea, H. N. *J. Chem. Educ.* **1969**, *46*, A844.

Brown, T. L.; LeMay, H. E., Jr.; Bursten, B. E. *Chemistry: The Central Science*; Prentice Hall: Englewood Cliffs, NJ, 1991; pp 467–471.

Kingzett, C. T. *Chemical Encyclopedia;* Van Nostrand: New York, 1928; pp 179–180.

Mellor, J. W. *A Comprehensive Treatise on Inorganic and Theoretical Chemistry;* Longmans, Green & Co.: London, 1923; Vol. 3, pp 554–564.

Nebergall, W. H.; Schmidt, F. C.; Holtzclaw, H.F., Jr. *College Chemistry*, 5th ed.; Heath: Lexington, MA, 1976; pp 317–322.

Purple of Cassius

Heating a colorless solution of glucose mixed with a colorless solution of hydrogen tetrachloroaurate(III) results in the formation of purple of Cassius.

Procedure

Please consult the Safety Information before proceeding. Use a microtip pipet when dispensing drops.

1. Heat about 150 mL of water to boiling in a 250-mL beaker.
2. Place 5 mL of glucose solution in a test tube.
3. Add 2 drops of 0.1% hydrogen tetrachloroaurate(III). Mix the solutions thoroughly.
4. Place the test tube in the boiling water bath and heat for 2 min. Heating will result in purple of Cassius.

Safety Information

1. Sodium hydroxide is corrosive to all tissues. Inhalation of the dust or concentrated mist may cause damage to the respiratory tract. Wear gloves, an apron, and goggles when handling.

2. Although glucose is a food substance, no substance should ever be tasted in the laboratory.

3. Hydrogen tetrachloroaurate(III) is harmful if swallowed, inhaled, or absorbed through the skin. The material is extremely destructive to the tissue of mucous membranes and the upper respiratory tract, eyes, and skin. It should be used only in a fume hood.

Materials

▲ Hydrogen tetrachloro-aurate(III) trihydrate, 0.1% $HAuCl_4 \cdot 3H_2O$: Dissolve 0.1 g in enough distilled water to make 100 mL of solution. Store the solution stoppered and in the dark. Light will affect the solution.

▲ Glucose, 0.5 M $C_6H_{12}O_6$: Dissolve 9.0 g in enough distilled water to make 100 mL of solution.

▲ Sodium hydroxide, 1.0 M NaOH: Dissolve 4.0 g in enough distilled water to make 100 mL of solution.

▲ Helium–neon laser light source.

5. Place the test tube in front of a helium–neon laser beam. The scattered red beam will be easily seen. Exercise extreme caution when using any laser and avoid all direct contact with the eyes.

Concepts

▲ The production of purple of Cassius is mentioned by Kingzett as occurring when potassium carbonate is added to gold chloride solution, the mixture is heated, and then glucose solution is added. With continued heating, the mixture becomes a brilliant purple, known as purple of Cassius.

▲ Purple of Cassius is often obtained by adding a solution of tin(II) chloride to dilute gold chloride. It is assumed that gold is reduced and tin(IV) chloride is formed. The precipitate formed may be various shades of brown, purple, blue, green, yellow, or red.

▲ Mellor noted that purple of Cassius behaves like a typical colloid toward an electric current and on dialysis.

▲ Colloids are discussed in the Concepts section of Investigation 32.

▲ Various arguments have been proposed to explain the composition of purple of Cassius, but there is no agreement. When tin(II) chloride is used, it is suggested that the different varieties of stannic acid explain the contradictions concerning purple of Cassius.

▲ A pamphlet was written by A. Cassius about 1684 entitled *De Auro* (Leiden, 1685). It described, for the first time, the preparation of a purple color by mixing tin and gold chlorides.

Reactions

Glucose reduces the gold(III) ion, and colloidal gold forms.

Notes

1. If a helium–neon laser is not available, place a piece of paper with a small hole cut in the middle over an overhead projector light so you will have a more direct beam of light that can be observed as it is scattered by the colloid.

2. The mixture will test as a colloid when first made, but solid will settle out after a day. Colloidal-sized particles sometime aggregate because of partial charges, forming clusters too large to remain suspended.

References

Brown, T. L.; Le May, H. E., Jr.; Bursten, B. E. *Chemistry: The Central Science;* Prentice Hall: Englewood Cliffs, NJ, 1991; pp 467–471.

Kingzett, C. T. *Chemical Encyclopedia*, 4th ed.; Van Nostrand: New York, 1928; pp 180, 588.

Mellor, J. W. *A Comprehensive Treatise on Inorganic and Theoretical Chemistry;* Longmans, Green & Co.: London, 1923; Vol. 3, pp 564–567.

Measurement of Current during Titration

The current, in milliamperes, in a solution of silver acetate is recorded, and the light emitted by the filaments of a light-emitting diode (LED) apparatus is noted. As hydrochloric acid is added to the solution, the current is recorded. A precipitate forms during the addition of the acid. During the titration, the current reading gets lower and the LED grows progressively dimmer during the titration and finally goes out. With continued addition of acid, the current increases, and the LED grows brighter again.

Procedure

Please consult the Safety Information before proceeding.

1. Using the graduations on a 250-mL beaker, add 150 mL of distilled water. Place the beaker on a magnetic stirrer. Using a graduated cylinder, add 30 mL of silver acetate solution. Turn on

Safety Information

1. External contact with hydrochloric acid causes severe burns, and contact with the eyes may cause permanent damage. Inhalation causes coughing and choking with possible inflammation and ulceration of the respiratory tract. Work in a fume hood and wear goggles, gloves, and a rubber apron.

2. Silver acetate may be harmful by inhalation, ingestion, or skin absorption and causes eye and skin irritation. It is irritating to mucous membranes and the upper respiratory tract.

Materials

▲   Silver acetate, $AgC_2H_3O_2$:
Add 83 g of solid to 1000 mL
of distilled water. Place on a
magnetic stirrer, mixing for
20 min. Filter the solution,
keeping the filtrate. Dry and
save the solid.

▲   Hydrochloric acid, 1.0 M
HCl: Adding acid to water,
mix 83 mL of concentrated
acid with enough water to
make 1000 mL of solution.

▲   Multimeter.

▲   LED apparatus.

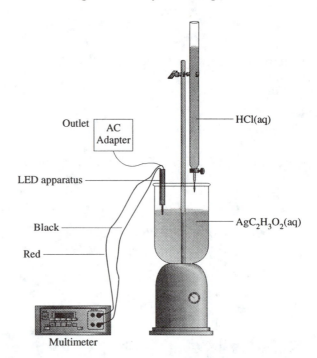

the stirrer at high speed to ensure thorough
mixing.

2.   Add 20 mL of hydrochloric acid to the buret,
and position the buret so the acid can drip into
the beaker.

3.   Using a multimeter, place the red patch cord in
"A" and the black one in "COM". Using alligator clips, attach the red patch cord to the red
wire of the LED apparatus (consult Investigation 77) and attach the black patch cord to the
black wire of the LED apparatus.

4.   Turn the multimeter to 200 mA. If necessary,
press the DC/AC button so the screen shows
"AC".

5.   Plug in the adapter.

6.   Drape the wire of the LED apparatus over a
clamp on the ring stand with the electrodes in
the solution so the initial current on the multimeter reads close to 16.7 mA. Note the brightness of the LED and record the initial current
of the silver acetate solution.

7.   Dropwise, with stirring, add hydrochloric acid
until the LED is extinguished. Have students
record observations and current readings ini-

tially every 2–3 drops and then for every drop after 1.80 mL has been added.

8. Continue to add hydrochloric acid until the LED is again lit.

Concepts

▲ For a solution to conduct a current, ions have to be present. When solutions of silver acetate and hydrochloric acid are made, silver, acetate, hydronium, and chloride ions are formed by dissociation of the compounds.

▲ Conductance decreases when the concentration of ions decreases because removal of ions via formation of a precipitate, silver chloride, occurs.

▲ Additional acid then increases the conductance of the solution as more hydronium and chloride ions are formed.

▲ Using the apparatus described and solutions, we obtained these results:

  • initial current, 16.7 mA

  • volume when LED is extinguished, 1.91 mL

  • current when LED is extinguished, 4.5 mA

▲ Ion-exchange resins are used to soften and deionize water. Two ion-exchange resins are used to produce high-quality, almost ion-free water. Present are both an acidic cation-exchange resin, which removes positively charged ions by replacing them with H^+ ions, and a basic anion-exchange resin to remove negatively charged ions, replacing them with OH^- ions. The result is the formation of water. The effectiveness of this process is tested with a conductivity meter. When no current is conducted, the ion exchange has taken place and water with no ions has been formed.

Reactions

1. $AgC_2H_3O_2(aq) \rightarrow Ag^+(aq) + C_2H_3O_2^-(aq)$
2. $HCl(aq) + H_2O(l) \rightarrow H_3O^+(aq) + Cl^-(aq)$
3. $Ag^+(aq) + Cl^-(aq) \leftrightarrows AgCl(s)$

Notes

1. To make this demonstration visually more interesting, add universal indicator to the solution in the beaker. The indicator proceeds through the spectrum of colors from red to blue. Some of the colors flash through the solution just before the solution takes on that color.

2. Barium hydroxide and 1.0 M sulfuric acid can also be used, but then the initial current of the barium hydroxide should be around 31.2 mA.

3. Students might question the solution of silver acetate because it has low solubility. Explain to them how the solution was made.

References

Gunn, E. L. *J. Chem. Educ.* **1940**, *17*, 385.

McClelland, F. *J. Chem. Educ.* **1930**, *7*, 1579.

McQuarrie, D. A.; Rock, P. A. *General Chemistry;* Freeman: New York, 1987; pp 538–540.

Stone, H. W. *J. Chem. Educ.* **1930**, *7*, 2722.

Determination of Solubility Product Constants Using Current

By using compounds that have a solubility product constant (K_{sp}) around 10^{-3}–10^{-7}, an approximate value for K_{sp} can be determined by comparing the current in a solution of the compound to the current in a known concentration of hydrochloric acid (see Investigation 77).

Procedure

Please consult the Safety Information before proceeding.

1. Using a multimeter, place a red patch cord in "A" and a black one in "COM". Using an alligator clip and a wire, attach the red patch cord to the LED apparatus (consult Investigation 77) and attach the black patch cord to the LED apparatus using another alligator clip and wire.

2. Set the multimeter on 200 mA. If necessary,

Safety Information

1. Calcium sulfate and lead chloride may be harmful if inhaled or swallowed. They may cause eye and skin irritation.

2. Silver sulfate may be harmful if inhaled, ingested, or absorbed through the skin. It causes eye and skin irritation and is irritating to mucous membranes and the upper respiratory tract.

3. Calcium hydroxide is harmful if swallowed, inhaled, or absorbed through the skin. The material is extremely destructive to the tissue of the mucous membranes and the upper respiratory tract, eyes, and skin. →

4. Barium hydroxide octa-hydrate may be fatal if inhaled, swallowed, or absorbed through the skin. The material is extremely destructive to the tissue of the mucous membranes and the upper respiratory tract, eyes, and skin.

press the DC/AC button so the screen shows "AC".

3. Place 40 mL of solution in a 50-mL beaker.

4. Plug in the adapter.

5. Insert the probes of the LED apparatus in the solution for 10 s to a specific height (see Notes). Record the current.

6. Rinse the probes with distilled water and wipe them dry. Repeat the reading two more times, washing and drying the electrodes between trials. Average the results of the three trials.

7. Repeat Steps 3–6 using the other solutions.

8. Using the graph of concentration versus current from Investigation 77, determine the concentration of HCl that corresponds to the current value for the solution. Calculate K_{sp}.

Concepts

▲   Some compounds are only slightly soluble (essentially insoluble) in water. When the compound is dissolved in water, an equilibrium exists between undissolved solid and ions in solution.

▲   The solubility product constant, K_{sp}, can be calculated if the concentration of ions in solution is known. Following are two examples and information that illustrates how to calculate the concentration and K_{sp}:

Compound	Current (mA)	Concentration of HCl with the Same Current (M)
Ag_2SO_4	27.4	0.0148
$CaSO_4$	22.3	0.00775

Hydrochloric acid would produce 2 mol of ions, so the total concentration for silver sulfate is 2 × 0.0148 M = 0.0296 M. Two-thirds of the total is represented by silver ions, which is (2/3) × 0.0296 M = 0.0197 M. One-third of the total is represented by sulfate ions, which is (1/3) × 0.0296 M = 0.00987 M.

K_{sp} for $Ag_2SO_4 = [Ag^+]^2[SO_4^{2-}] = (0.0197$ M$)^2(0.00987$ M$) = 3.80 \times 10^{-6}$

The total concentration of calcium sulfate is 2×0.00775 M $= 0.0155$ M.

One-half of the total is represented by calcium ions, which is $(1/2) \times 0.0155 = 0.00775$. The concentration of sulfate ions is also $(1/2) \times 0.0155$ M $= 0.00775$ M.

K_{sp} for $CaSO_4 = [Ca^{2+}][SO_4^{2-}] = (0.00775$ M$)(0.00775$ M$) = 6.01 \times 10^{-5}$

▲ Calculating K_{sp} with only molarities does not take into account interactions of ions with each other and with solvent molecules. Ion pairs exist between the cation and anion, such as Ca^{2+} and OH^-, forming $CaOH^-$. These pairs reduce the mobility of each ion in an electric field and affect the stoichiometric concentration. The *activity* of a solution takes into account interionic attractions. Relating the activity of a solution to its stoichiometric concentration through a factor called the *activity coefficient* would allow a more accurate solubility product constant to be calculated.

▲ K_{sp} values at 25 °C, as found in *Lange's Handbook of Chemistry*, and experimental values based on the results are as follows:

Compound	K_{sp}	Experimental
Calcium sulfate	9.1×10^{-6}	6.01×10^{-5}
Silver sulfate	1.4×10^{-5}	3.80×10^{-6}
Calcium hydroxide	5.5×10^{-6}	4.40×10^{-6}
Lead chloride	1.6×10^{-5}	5.30×10^{-6}
Barium hydroxide	5.0×10^{-3}	7.26×10^{-3}

▲ Following are the other three compounds, the current, and concentration of HCl with the same current based on the device used:

Compound	Current (mA)	Concentration of HCl with the Same Current (M)
$Ca(OH)_2$	27.7	0.0155
$PbCl_2$	28.1	0.0165
$Ba(OH)_2$	34.3	0.183

Materials

▲ All solutions—$CaSO_4$, Ag_2SO_4, $Ca(OH)_2$, $PbCl_2$, and $Ba(OH)_2$:

1. Add an excess of solid to 200 mL of distilled water. Because the solids do not dissolve to any appreciable extent, it will be obvious when there is an excess.

2. Place on a magnetic stirrer and stir for at least 20 min.

3. Filter and retain the filtrate.

4. Store in a labeled bottle.

▲ Multimeter.

▲ LED apparatus.

Reactions

$$CaSO_4(s) \leftrightharpoons Ca^{2+}(aq) + SO_4^{2-}(aq)$$
$$Ag_2SO_4(s) \leftrightharpoons 2Ag^+(aq) + SO_4^{2-}(aq)$$
$$Ca(OH)_2(s) \leftrightharpoons Ca^{2+}(aq) + 2OH^-(aq)$$
$$PbCl_2(s) \leftrightharpoons Pb^{2+}(aq) + 2Cl^-(aq)$$
$$Ba(OH)_2(s) \leftrightharpoons Ba^{2+}(aq) + 2OH^-(aq)$$

Notes

1. It is important to filter the solutions before using, as opposed to simply using the "solution" after the excess solid has settled out. K_{sp} is closer to the correct value when the interference by solid is removed.

2. The choice of 40 mL of solution and a 50-mL beaker is somewhat arbitrary. What is necessary, though, is that the same volume of solution in the same-size beaker be used each time and the probes submerged the same distance (consult Notes in Investigation 77).

3. The solutions can be used over and over again as long as the probes are rinsed and dried before using in another solution. Save the solutions in labeled bottles.

4. The experimental results are not meant to agree quantitatively with results found in *Lange's Handbook of Chemistry*, but the demonstration is meant to provide a means by which similar results can be obtained and K_{sp} can then be calculated by your students.

References

Dean, J. A.; *Lange's Handbook of Chemistry,* 14th ed.; McGraw-Hill: New York, 1992; pp 8.6–8.11.

Krauss, C. A. *J. Chem. Educ.* **1958**, *35*, 324.

Petrucci, R. *General Chemistry*, 5th ed.; Macmillan: New York, 1989; pp 462, 686–687.

Decreasing Solubility at a Higher Temperature

When a solution of sodium sulfite is heated, the solute undergoes a decrease in solubility and sodium sulfite solid appears in solution. Upon chilling the solution, the solid dissolves.

Procedure

Please consult the Safety Information before proceeding.

1. Place the prepared sodium sulfite solution in a loosely stoppered flask on a hot plate–stirrer. Bring the solution to a vigorous boil. Solid will begin to appear as a film on the surface. When this reaction occurs, turn on the stirrer so the solid can be observed moving through the solution like snowflakes.

2. Remove the flask from the heat and place it in an ice bath on a magnetic stirrer. Cool the contents and stir to dissolve the solid.

Safety Information

Anhydrous sodium sulfite is moderately toxic by ingestion. If inhaled, it may cause mucous membrane irritation. Skin contact may cause irritation and contact dermatitis.

Materials

▲ Sodium sulfite: Place 200 mL of water in a beaker on a hot plate–stirrer. Turn on the stirrer and hot plate and heat the water to about 60 °C. Continue heating the water at about the same temperature and add 40 g of anhydrous sodium sulfite, Na_2SO_3. When all the solid has dissolved, add 5 more grams. Heat 2–3 min, remove the mixture from the heat, decant the solution, and store the solution in a stoppered flask.

3. Repeat Steps 1 and 2 several times.

Concepts

▲ Solubility is defined as the amount of solute (grams) that will dissolve in 100 mL of water.

▲ Solubility usually increases as temperature increases and decreases as temperature decreases.

▲ Some solids, such as calcium acetate, cerium(III) sulfate, and manganese(II) sulfate, have a negative coefficient of solubility. When the temperature of the solution is increased, instead of more solid dissolving, solid starts to precipitate out of solution. Likewise, as the temperature is lowered, the excess solid dissolves.

Reactions

$$2Na^+(aq) + SO_3^{2-}(aq) \overset{\Delta}{\rightarrow} Na_2SO_3(s)$$

Notes

1. For the preparation of other substances demonstrating a negative coefficient of solubility, consult any of the References.

2. Before heating the flask for the demonstration, mark the level of solution in the flask. After completing the demonstration, replenish water to this mark if any evaporation takes place.

References

Bacon, E. K. *J. Chem. Educ.* **1938**, *15*, 494.

Bateman, C. A.; Fernelius, W. C. *J. Chem. Educ.* **1937**, *14*, 315.

Kobe, K. A. *J. Chem. Educ.* **1959**, *16*, 183.

Sorum, C. H. *J. Chem. Educ.* **1928**, *5*, 1287.

Formation of Red, White, and Blue Precipitates

Adding a colorless solution of iron(II) ammonium sulfate to three different beakers results in the slow production of red, white, and blue precipitates.

Procedure

Please consult the Safety Information before proceeding.

1. In one 250-mL beaker, dissolve 2 g of sodium salicylate, $NaC_7H_5O_3$, in a small amount of water.

2. In a second 250-mL beaker, place the strontium chloride solution.

3. In a third 250-mL beaker, dissolve 2 g of potassium hexacyanoferrate(II) trihydrate, $K_4Fe(CN)_6 \cdot 3H_2O$, in a small amount of water.

Materials

▲ Strontium chloride: Dissolve 5 tsp (4 g) of strontium chloride in 25 mL of water. The process is endothermic. Let the solution warm to room temperature before using it.

▲ Sodium salicylate, powder.

▲ Potassium hexacyanoferrate(II) trihydrate, powder.

▲ Iron(II) ammonium sulfate hexahydrate, powder.

4. When ready to begin, dissolve 1.5 g of iron(II) ammonium sulfate hexahydrate, $Fe(NH_4)_2(SO_4)_2 \cdot 6H_2O$, in approximately 800 mL of water.

5. Pour equal amounts of solution from the large beaker into each of the 250-mL beakers. Stir each solution quickly after the addition.

6. Observe the slow production of red, white, and blue precipitates in the three different beakers.

Concepts

▲ Precipitates form in each of the three beakers and are responsible for the observed colors.

▲ The slow-developing colors are a result of the iron(II) ion, Fe^{2+}, being oxidized in air to the iron(III) ion, Fe^{3+}. Once this oxidation occurs, the full effect of the color can be observed.

▲ Iron(II) ammonium sulfate hexahydrate, which contributes iron(II) ions, is also known as Mohr's salt. Susceptibility of Fe^{2+} to aerial oxidation is dependent on the nature of the ligands attached. Alkaline solutions are very readily oxidized, whereas acid solutions are much more stable.

▲ The blue precipitate formed is called Prussian blue, which is formed when Fe(III) is added to hexacyanoferrate(II). Turnbull's blue is formed when Fe(II) is added to hexacyanoferrate(III). The distinction between the two is artificial, because the compound formed is the same for both of them.

▲ Discovery of the "blue" compounds took place in 1704 when they were used in the manufacture of inks and paints. They were then used in blueprints in 1840.

▲ When a small trace of iron(III) salts is present, salicylic acid and its salts are colored reddish; hence red iron(III) salicylate.

▲ The combination of aqueous Sr^{2+} and SO_4^{2-} results in the formation of the white precipitate, $SrSO_4$.

Reactions

$$3C_6H_4OHCOO^-(aq) + Fe^{3+}(aq) \leftrightarrows$$
$$Fe(C_6H_4OHCOO)_3(s)$$
$$\text{red}$$

$$Sr^{2+}(aq) + SO_4^{2-}(aq) \leftrightarrows SrSO_4(s)$$
$$\text{white}$$

Formation of Prussian blue:
$$Fe^{2+}(aq) + 6CN^-(aq) \rightarrow [Fe(CN)_6]^{4-}(aq)$$
$$\text{hexacyanoferrate(II)}$$

$$Fe^{3+}(aq) + K^+(aq) + [Fe(CN)_6]^{4-}(aq) + H_2O(l) \rightarrow$$
$$KFe[Fe(CN)_6] \cdot H_2O(s)$$
$$\text{Prussian blue}$$

Notes

1. Although this reaction can be done using iron(III) ammonium sulfate, it works much better with iron(II) ammonium sulfate.

2. Because iron(II) ammonium sulfate is oxidized in air to the iron(III) salt, the solution cannot be prepared in advance. The process of oxidation can be slowed down by placing cleaned iron filings in the iron(II) solution. To clean the filings, rinse them in ethanol and discard the alcohol. Repeat the rinsing two more times. Rinse once in acetone. Dry the filings in air before placing them in the iron(II) solution.

3. Give the reaction time to occur. The colors develop slowly.

4. Use the least amount of water needed to dissolve the sodium salicylate and potassium hexacyanoferrate(II) trihydrate.

References

Greenwood, N. N.; Earnshaw, A. *Chemistry of the Elements;* Pergamon: New York, 1984; pp 1269, 1271.

Hansen, L. D.; Litchman, W. M.; Daub, G. H. *J. Chem. Educ.* **1969,** *46,* 46.

Kohn, M. *J. Chem. Educ.* **1943**, *20,* 198.

Linstromberg, W. W.; Baumgarten, H. E. *Organic Chemistry;* Heath: Lexington, MA, 1978; p 372.

Lippy, J. D., Jr.; Palder, E. L. *Modern Chemical Magic,* Stackpole Co.: Harrisburg, PA, 1959; p 3.

The Merck Index, 10th ed.; Windholz, M., Ed.; Merck: Rahway, NJ, 1983; p 1200.

Nebergall, W. H.; Schmidt, F. C.; Holtzclaw, H. F., Jr. *College Chemistry*, 5th ed.; Heath: Lexington, MA, 1976; p 909.

Robin, M. B. *Inorg. Chem.* **1962**, *1*, 337.

Shriver, D. F.; Shriver, S. A.; Anderson, S. E. *Inorg. Chem.* **1965**, *4*, 725.

Colorful Beakers

Adding water to a series of beakers, each of which contains two substances, results in various colors in the beakers—red, lavender, yellow, green–blue, blue, and green.

Procedure

Please consult the Safety Information before proceeding.

1. Arrange six 250-mL beakers in a row. *Except for using 1 g of cobalt(II) chloride hexahydrate in beaker 6, place 0.5 g of each of the solids in the appropriate beaker as follows:*

   Beaker 1: Sodium salicylate and iron(II) ammonium sulfate hexahydrate

   Beaker 2: Cobalt(II) chloride hexahydrate and potassium carbonate

   Beaker 3: Calcium oxide and a few drops of methyl orange

4. Iron(II) ammonium sulfate hexahydrate may be harmful by inhalation, ingestion, or skin absorption. It causes eye and skin irritation. It is irritating to mucous membranes and the upper respiratory tract.

5. Potassium hexacyanoferrate(II) trihydrate is an eye and skin irritant. It does not appear to release the cyanide ion.

6. Potassium carbonate is an irritant through inhalation and eye and skin contact.

Beaker 4: Iron(II) ammonium sulfate hexahydrate and potassium carbonate

Beaker 5: Potassium hexacyanoferrate(II) trihydrate and iron(II) ammonium sulfate hexahydrate

Beaker 6: Cobalt(II) chloride hexahydrate and potassium hexacyanoferrate(II) trihydrate.

2. Fill each beaker about three-fourths full with water from a large pitcher or 1-L beaker, stirring quickly.

Concepts

▲ The formation of precipitates, except with calcium oxide, is responsible for the observed colors.

▲ When a small trace of iron(III) salts is present, salicylic acid and its salts are colored reddish, hence red iron(III) salicylate.

▲ Combination of Co^{2+} and CO_3^{2-} forms the lavender precipitate $CoCO_3(s)$, cobalt(II) carbonate.

▲ Calcium oxide combines with water to form calcium hydroxide, and this results in a basic solution. Methyl orange indicator changes in the pH range 3.1–4.4 and goes from red to yellow in a more basic solution.

▲ When the concentration of hydroxide ion, OH^- (aq), increases with the formation of calcium hydroxide, the equilibrium shifts to the product side of the indicator reaction, where methyl orange is yellow.

▲ A green–blue precipitate, iron (II) carbonate, forms between Fe^{2+} and CO_3^{2-}.

▲ A complex ion forms between Fe^{2+} and CN^- called hexacyanoferrate(II), $[Fe(CN)_6]^{4-}$. Fe^{2+} is oxidized to Fe^{3+} by oxygen in the air, and Fe^{3+}, K^+, and H_2O combine with the hexacyanoferrate(II) ion, forming the blue compound called Prussian blue. Since its discovery in 1704, Prussian blue has been used as a pigment in inks and paints. In 1840, it was used in the production of blueprints.

▲ The compound formed between cobalt(II) chloride and potassium hexacyanoferrate(II) is probably a combination of two compounds—

blue cobalt(II) cyanide, $Co(CN)_2$, and yellow potassium hexacyanocobaltate(III), $K_3[Co(CN)_6]$.

▲ If more potassium hexacyanoferrate(II) solid than cobalt(II) chloride hexahydrate solid is mixed with water, a yellow–green gelatinous mixture results. More potassium hexacyanocobaltate(III), which is freely soluble in water, than cobalt(II) cyanide is present. At the same time, some cobalt(II) cyanide, which is not soluble in water, is probably also formed. The addition of more cobalt(II) chloride hexahydrate than potassium hexacyanoferrate(II) in beaker 6 produces a green gelatinous precipitate. More of the blue cobalt(II) cyanide than potassium hexacyanocobaltate(III) is probably now present.

Materials

▲ Methyl orange: Dissolve 0.1 g in enough water to make 100 mL of solution.

▲ Sodium salicylate, powder.

▲ Cobalt(II) chloride hexahydrate, powder.

▲ Calcium oxide, powder.

▲ Iron(II) ammonium sulfate hexahydrate, powder.

▲ Potassium hexacyanoferrate(II) trihydrate.

▲ Potassium carbonate, powder.

Reactions

1. $Fe^{3+}(aq) + 3C_6H_4OHCOO^-(aq) \rightarrow$
$$Fe(C_6H_4OHCOO)_3(s)$$
red

2. $Co^{2+}(aq) + CO_3{}^{2-}(aq) \rightarrow CoCO_3(s)$
lavender

3. $CaO(s) + H_2O(l) \rightarrow Ca(OH)_2(s)$

4. General indicator reaction:
$HIn(aq) + OH^-(aq) \leftrightharpoons H_2O(l) + In^-(aq)$
red   yellow

5. Specific indicator reaction with methyl orange:

6. $Fe^{2+}(aq) + CO_3{}^{2-}(aq) \rightarrow FeCO_3(s)$
green–blue

7. Formation of Prussian blue:
$Fe^{2+}(aq) + 6CN^-(aq) \rightarrow [Fe(CN)_6]^{4-}(aq)$
hexacyanoferrate(II)

$$Fe^{3+}(aq) + K^+(aq) + [Fe(CN)_6]^{4-}(aq) + H_2O(l) \rightarrow$$
$$KFe[Fe(CN)_6]\cdot H_2O(s)$$
$$\text{Prussian blue}$$

8. Formation of combination cobalt compounds:

$$Co^{2+}(aq) + 2CN^-(aq) \rightarrow Co(CN)_2(s)$$
$$\text{blue}$$

$$3K^+(xaq) + [Co(CN)_6]^{3-}(aq) \rightarrow K_3[Co(CN)_6](s)$$
$$\text{yellow}$$

Notes

1. Make sure to use twice as much cobalt(II) chloride hexahydrate as potassium hexacyanoferrate(II) in beaker 6.

2. Other indicators could be used with the calcium oxide for different color effects, such an phenolphthalein (pink), bromocresol purple (purple), or phenol red (red).

3. Sodium hexacyanoferrate(II) can be used in place of potassium hexacyanoferrate(II).

4. Although iron(III) ammonium sulfate could be used in place of iron(II) ammonium sulfate, we prefer the iron(II) salt.

5. Use your imagination and add other colors.

References

Greenwood, N. N.; Earnshaw, A. *Chemistry of the Elements;* Pergamon: New York, 1984; pp 1271–1272, 1301–1302, 1306, 1308, 1311, 1315–1316.

Hansen, L. D.; Litchman, W. M.; Daub, G. H. *J. Chem. Educ.* **1969,** *46,* 46.

Kohn, M. *J. Chem. Educ.* **1943,** *20,* 198.

Lippy, J. D., Jr.; Palder, E. L. *Modern Chemical Magic;* Stackpole Co.: Harrisburg, PA, 1959; p 5.

The Merck Index, 10th ed.; Windholz, M., Ed.; Merck: Rahway, NJ, 1983; pp 346, 7519.

Nebergall, W. H.; Schmidt, F. C.; Holtzclaw, H. F., Jr. *College Chemistry,* 5th ed.; Heath: Lexington, MA, 1976; p 909.

Shakhashiri, B. Z. *Chemical Demonstrations: A Handbook for Teachers of Chemistry;* University of Wisconsin: Madison, WI, 1989; Vol. 3, p 23.

Interstitial Compounds

When barium sulfate is precipitated in the presence of potassium permanganate, some of the potassium permanganate is trapped inside the barium sulfate crystals.

Procedure

Please consult the Safety Information before proceeding.

1. Place 5 mL of barium hydroxide solution in each of two test tubes.

2. Add 5 drops of potassium permanganate solution to the first test tube. Mix the solution thoroughly.

3. Add drops of 6 M sulfuric acid solution to both test tubes until a heavy precipitate forms. The precipitate will appear pale purple in the first test tube and white in the second test tube. Thoroughly mix the contents of each test tube.

3. Sodium thiosulfate is a strong reducing agent and may be irritating through skin contact.

4. Barium hydroxide is extremely corrosive to body tissue. Handle with care and avoid all contact with the skin. Inhalation of the dust results in corrosive action on the mucous membranes. Wear gloves when handling. Barium compounds are poisonous.

4. Centrifuge both test tubes or let them sit for 5 min. Decant the solution and wash the precipitate in the first test tube several times until the supernatant liquid is colorless. Discard the supernatant liquid.

5. Place 5 mL of distilled water in a third test tube. Add 5 drops of potassium permanganate solution and mix the solution thoroughly.

6. Add 1 mL of sodium thiosulfate solution to the first and third test tubes. Mix each thoroughly and observe the color change in the third test tube but not in the first test tube.

Concepts

▲ Barium sulfate is insoluble in water and therefore precipitates when sulfuric acid and barium hydroxide are allowed to react. Barium sulfate is very dense and settles to the bottom of the test tube very quickly.

▲ When permanganate ions are present with precipitating barium sulfate, permanganate ions are trapped in the forming barium sulfate crystals.

▲ When thiosulfate ions come in contact with permanganate ions, a redox reaction takes place. Purple permanganate ions are reduced to manganese(II) ions, which do not exhibit color, and thiosulfate ions are oxidized to sulfate. When permanganate ions are trapped within the barium sulfate crystal, thiosulfate ions are unable to react with them.

Reactions

1. $Ba^{2+}(aq) + SO_4^{2-}(aq) \rightarrow BaSO_4(s)$

2. $Ba^{2+}(aq) + SO_4^{2-}(aq) + MnO_4^-(aq) \rightarrow$
$$[(BaSO_4)(MnO_4^-)] (s)$$

3. $14H^+(aq) + 8MnO_4^-(aq) + 5S_2O_3^{2-}(aq) \rightarrow$
$$8Mn^{2+}(aq) + 10SO_4^{2-}(aq) + 7H_2O$$

Notes

1. Lead ions can be substituted for the barium ions to precipitate lead sulfate. Strontium sulfate will not behave in a similar manner, and this fact can be used to detect as little as 1 ppm of barium in the presence of strontium.

2. As a variation, several drops of potassium permanganate solution can be added to the second tube—the pure precipitated barium sulfate—to show that the permanganate ions are not adsorbed.

References

Feigl, F. *Spot Tests, Vol. I, Inorganic Chemistry*, 4th English translation; Elsevier: Houston, TX, 1954; p 205.

Materials

▲ Sulfuric acid, 6.0 M H_2SO_4: Add 33.6 mL of concentrated acid solution to enough water to make 100 mL of solution.

▲ Barium hydroxide dihydrate, 0.5 M $Ba(OH)_2 \cdot 2H_2O$: Dissolve 15 g in enough water to make 100 mL of solution.

▲ Sodium thiosulfate pentahydrate, 1.0 M $Na_2S_2O_3 \cdot 5H_2O$: Dissolve 24 g in enough water to make 100 mL of solution.

▲ Potassium permanganate, 0.01 M $KMnO_4$: Dissolve 1.6 g in enough water to make 100 mL of solution.

Transition Metals and Complex Ions

Nitrate Ring

Equal amounts of iron(II) sulfate solution and a nitrate solution are mixed and shaken in a test tube. A pipet containing concentrated sulfuric acid is carefully placed with its tip underneath the solutions, and the acid is released. A brown ring forms at the interface.

Procedure

Please consult the Safety Information before proceeding.

1. Place 5 mL of iron(II) sulfate solution in a small test tube.

2. Add an equal amount of any of the nitrate solutions, stopper the tube, and shake the tube vigorously.

3. Fill a microtip pipet full with concentrated sulfuric acid.

Safety Information

1. Iron(II) sulfate heptahydrate may be harmful by inhalation, ingestion, or skin absorption. It causes eye and skin irritation. It is irritating to mucous membranes and the upper respiratory tract.

2. Concentrated sulfuric acid is very corrosive. Handle with caution and avoid contact with the skin because it produces severe burns. Inhalation of the concentrated vapor may cause serious lung damage. Work in a fume hood and wear goggles, gloves, and a rubber apron. →

3. Sodium and calcium nitrate may be harmful by inhalation, ingestion, or skin absorption. They cause eye and skin irritation. They are irritating to mucous membranes and the upper respiratory tract.

4. Zinc nitrate is harmful if swallowed, inhaled, or absorbed through the skin. The material is extremely destructive to the tissue of the mucous membranes and the upper respiratory tract, eyes, and skin.

4. Rinse any excess acid off the pipet.

5. Carefully place the pipet tip underneath the solution at the bottom of the test tube.

6. Slowly release the acid. Continuing to squeeze the bulb, remove the pipet from the test tube.

7. A brown ring will develop at the interface.

8. Repeat with other nitrate solutions.

Concepts

▲ Fe^{2+} is oxidized to Fe^{3+} through the loss of electrons. Because Fe^{2+} is oxidized, it acts as the reducing agent. The half-reaction is

$$3(Fe^{2+} \rightarrow Fe^{3+} + e^-)$$

▲ Reduction of nitrogen takes place by gaining electrons. The overall charge on the NO_3 ion is 1–. The three oxygens each have a charge of 2–, which equals a total negative charge of 6–. The charge on the nitrogen is therefore 5+. The charge on nitrogen in NO is 2+. Because nitrogen is reduced, it acts as the oxidizing agent according to

$$NO_3^- + 4H^+ + 3e^- \rightarrow NO + 2H_2O$$

▲ Very high H^+ ion concentration and relatively high temperatures at the interface of the two solutions cause the reaction to take place rapidly.

▲ The brown ring is a cationic iron nitrosyl complex that is $[Fe(NO)(H_2O)_5]^{2+}$.

▲ A number of principles are involved in this demonstration. Foremost is probably oxidation–reduction, followed by the formation of a complex ion, and then the observation of various layers in the test tube because of the different densities of the solutions.

Reactions

1. Oxidation of iron(II) to iron(III) causes some of the nitrate to be reduced to NO, nitric oxide.

$$4H^+(aq) + 3Fe^{2+}(aq) + NO_3^-(aq) \rightarrow$$
$$3Fe^{3+}(aq) + NO(aq) + 2H_2O(l)$$

2. The NO combines with iron(II) to form the nitrate ring.

$$Fe^{2+}(aq) + NO(aq) + 5H_2O(l) \rightarrow$$
$$[Fe(NO)(H_2O)_5]^{2+}(aq)$$

Notes

1. It is important to rinse any excess acid off the pipet so the nitrate ring will form only at the interface.

2. The iron(II) sulfate solution must be prepared fresh. The Fe^{2+} is oxidized by the oxygen in the air, loses electrons, and becomes Fe^{3+}. You can use the solution throughout the day provided that you place about 1 Tbsp of clean, degreased iron filings in the solution and keep the beaker covered. To clean and degrease, rinse the iron filings three separate times in ethanol, pour off the ethanol each time. Rinse once in acetone. Air-dry the filings.

3. Observations about the different layers in the test tubes would be interesting to discuss with your students.

4. If any problems are encountered with the sulfuric acid, use acid from a new bottle.

Materials

▲ Sodium nitrate, 1.0 M $NaNO_3$: Dissolve 8.5 g in enough water to make 100 mL of solution.

▲ Calcium nitrate, 1.0 M $Ca(NO_3)_2$: Dissolve 23.6 g in enough water to make 100 mL of solution.

▲ Zinc nitrate, 1.0 M $Zn(NO_3)_2$: Dissolve 18.9 g in enough water to make 100 mL of solution.

▲ Iron(II) sulfate heptahydrate, 0.5 M $Fe(SO_4)_2 \cdot 7H_2O$: Dissolve 13.9 g in enough water to make 100 mL of solution.

References

Greenwood, N. N.; Earnshaw, A. *Chemistry of the Elements*, 1st ed.; Pergamon: New York, 1984; p 1272.

McAlpine, R. K.; Soule, B. A. *Fundamentals of Qualitative Analysis*, 4th ed.; Van Nostrand: New York, 1956; p 214.

Nebergall, W. H.; Schmidt, F. C.; Holtzclaw, H. F., Jr. *College Chemistry*, 5th ed.; Heath: Lexington, MA, 1976; p 1018.

Various Metal Ammine Complexes

When concentrated ammonia is added to different metallic nitrate solutions, precipitates form and color changes take place.

Procedure

Please consult the Safety Information before proceeding.

1. Place 10 mL of each of the six nitrate solutions in six different test tubes.

2. Without mixing, add 20 drops of concentrated ammonia to each test tube.

3. Have students record observations. Leave the test tubes out until the next day for additional observations.

4. Copper(II) nitrate hexahydrate is corrosive to eyes and skin. It is an irritant by inhalation and is moderately toxic by ingestion.

Concepts

▲ The metal hydroxides of the cobalt, nickel, copper, and zinc families are all insoluble in water. Aqueous ammonia reacts with the metal ions of these families to form insoluble metal hydroxides.

▲ The metal hydroxides will dissolve in an excess of aqueous ammonia to form ammine complexes.

▲ In this demonstration, if the ammonia is not mixed with the nitrate solutions, some of the colored complexes will be observed above a precipitate ring. The precipitate will eventually cascade down through the excess nitrate solution and form at the bottom of the test tube.

▲ In 1798, Tassaert accidentally discovered $[Co(NH_3)_6]Cl_3$, hexaamminecobalt(III) chloride. This is considered the first discovery of a metal complex.

▲ Alfred Werner, professor of chemistry in Zurich, and his associates prepared more than 700 cobalt coordination compounds. Their work increased chemical knowledge in this area and allowed Werner to propose his coordination theory in 1893 at the age of 26. But it was not until he had received the Nobel Prize in 1913 and after his death that his ideas were confirmed.

▲ Werner used the metal as a central element and, on the basis of his experimental work, surrounded it with six molecules (or ions for cobalt compounds). How many molecules or ions were attracted to the central metal determined the secondary valence, and the oxidation state of the metal was the primary valence. He termed the secondary valence the *coordination number*. If the six coordination spheres are treated in a geometric fashion, the metal is at the center of a regular octahedron. Werner extended his reasoning to hydrates, cyanides, thiocyanates, cyanates, carbonyls, and similar compounds.

Reactions

1. Formation of the insoluble silver hydroxide:

$Ag^+(aq) + NH_3(aq) + H_2O(l) \rightarrow$
$$AgOH(s) + NH_4^+(aq)$$

$$2AgOH(s) \rightarrow Ag_2O(s) + H_2O(l)$$

2. For Co^{2+}, Cu^{2+}, Zn^{2+}, Ni^{2+}, and Cd^{2+}, the same general reaction takes place (M is the metal ion):

$M^{2+}(aq) + 2NH_3(aq) + 2H_2O(l) \rightarrow$
$$M(OH)_2(s) + 2NH_4^+(aq)$$

3. Formation of the ammine complex:

$$AgOH(s) + 2NH_3(aq) \rightarrow [Ag(NH_3)_2]^+(aq) + OH^-(aq)$$

$$Co(OH)_2(s) + 6NH_3(aq) \rightarrow [Co(NH_3)_6]^{2+}(aq) + 2OH^-(aq)$$

$$Cu(OH)_2(s) + 4NH_3(aq) \rightarrow [Cu(NH_3)_4]^{2+}(aq) + 2OH^-(aq)$$

$$Zn(OH)_2(s) + 4NH_3(aq) \rightarrow [Zn(NH_3)_4]^{2+}(aq) + 2OH^-(aq)$$

$$Ni(OH)_2(s) + 6NH_3(aq) \rightarrow [Ni(NH_3)_6]^{2+}(aq) + 2OH^-(aq)$$

$$Cd(OH)_2(s) + 4NH_3(aq) \rightarrow [Cd(NH_3)_4]^{2+}(aq) + 2OH^-(aq)$$

Materials

▲ The following solutions are all 1.0 M. Dissolve the specified amount in enough water to make 100 mL of solution:

a. Silver nitrate, $AgNO_3$: 16.99 g.

b. Cobalt(II) nitrate hexahydrate, $Co(NO_3)_2 \cdot 6H_2O$: 29.10 g.

c. Copper(II) nitrate hemipentahydrate, $Cu(NO_3)_2 \cdot 2.5 H_2O$: 23.25 g.

d. Zinc nitrate heptahydrate, $Zn(NO_3)_2 \cdot 7H_2O$: 29.75 g.

e. Nickel(II) nitrate hexahydrate, $Ni(NO_3)_2 \cdot 6H_2O$: 29.09 g.

f. Cadmium nitrate tetrahydrate, $Cd(NO_3)_2 \cdot 4H_2O$: 30.85 g.

▲ Concentrated ammonia: commercial solution.

Notes

Mixing the ammonia with each nitrate solution will ultimately result in the formation of the ammine complex. Some of the solutions will undergo a color change, which indicates the presence of the complex. It is better, though, not to mix the ammonia but simply add it dropwise so both the complex when it is colored and the insoluble metal hydroxide can be observed.

References

Basolo, F.; Johnson, R. C. *Coordination Chemistry;* Benjamin: Menlo Park, CA, 1964; pp 3–12.

Werner Centennial; American Chemical Society: Washington, DC, 1967; pp 70–77.

Whitten, K. W.; Gailey, K. D. *General Chemistry with Qualitative Analysis*, 2nd ed.; Saunders: Philadelphia, PA, 1984; pp 797–798.

Rainbow Precipitates

Adding water to a series of beakers, each containing two solids, results in red, orange, yellow, green, blue, and violet precipitates.

Procedure

Please consult the Safety Information before proceeding.

1. In six 250-mL beakers, place 0.5 g (1/4 tsp) of each of the following solids.

    Beaker 1: Cobalt(II) chloride hexahydrate and potassium hexacyanoferrate(III)

    Beaker 2: Silver nitrate and potassium hexacyanoferrate(III)

    Beaker 3: Copper(II) chloride dihydrate and potassium hexacyanoferrate(III)

    Beaker 4: Nickel(II) nitrate and sodium hydroxide (crushed)

5. Sodium carbonate may be harmful by inhalation, ingestion, or skin absorption. It causes eye and skin irritation. It is irritating to mucous membranes and the upper respiratory tract.

6. Nickel(II) nitrate is an eye, skin, and mucous membrane irritant and is moderately toxic by ingestion.

7. Sodium hydroxide is corrosive to all tissues. Inhalation of the dust or concentrated mist may cause damage to the respiratory tract. Wear gloves, goggles, and a rubber apron when handling.

Beaker 5: Copper(II) chloride dihydrate and sodium carbonate

Beaker 6: Cobalt(II) chloride hexahydrate and potassium carbonate

2. Pouring from a large pitcher or beaker of water, fill each beaker about three-fourths full, stirring after the addition of the water. Colored precipitates of red, orange, yellow, green, blue, and violet are formed in beakers 1 through 6, respectively.

Concepts

▲   The colors, formulas, and names of the precipitates are

Color	Formula	Name
Red	$Co_3[Fe(CN)_6]_2$	Cobalt(II) hexacyanoferrate(III)
Orange	$Ag_3[Fe(CN)_6]$	Silver hexacyanoferrate(III)
Yellow	$Cu_3[Fe(CN)_6]_2 \cdot 14H_2O$	Copper(II) hexacyanoferrate(III) tetradecahydrate
Green	$Ni(OH)_2$	Nickel(II) hydroxide
Blue	$CuCO_3 \cdot Cu(OH)_2$	Copper(II) carbonate hydroxide
Violet	$CoCO_3 \cdot 6H_2O$	Cobalt(II) carbonate hexahydrate

▲   When two soluble ionic compounds are mixed, a double replacement reaction, or metathesis, sometimes takes place. The product of this reaction is an insoluble compound that precipitates out of solution.

▲   A net ionic reaction for metathesis involves only the ions that actually form the precipitate. The other ions that remain in solution are known as *spectator ions* and are not included in the net equation.

Reactions

1. $3Co^{2+}(aq) + 2[Fe(CN)_6]^{3-}(aq) \rightarrow Co_3[Fe(CN)_6]_2(s)$

2. $3Ag^+(aq) + [Fe(CN)_6]^{3-}(aq) \rightarrow Ag_3[Fe(CN)_6](s)$

3. $3Cu^{2+}(aq) + 2[Fe(CN)_6]^{3-}(aq) + 14H_2O(l) \rightarrow$
$$Cu_3[Fe(CN)_6]_2 \cdot 14H_2O(s)$$

4. $Ni^{2+}(aq) + 2OH^-(aq) \rightarrow Ni(OH)_2(s)$

5. $2Cu^{2+}(aq) + CO_3^{2-}(aq) + 2OH^-(aq) \rightarrow$
$$CuCO_3 \cdot Cu(OH)_2(s)$$

6. $Co^{2+}(aq) + CO_3^{2-}(aq) + 6H_2O(l) \rightarrow$
$$CoCO_3 \cdot 6H_2O(s)$$

Notes

1. Although 0.1 M solutions could be made for each compound, the demonstration is more interesting with just the solids. You are unable to "see" what is going to happen.

2. By using the *CRC Handbook of Chemistry and Physics*, other interesting combinations of colors can be created.

Reference

CRC Handbook of Chemistry and Physics, 72nd ed.; Lide, D. R., Ed.; CRC: Boca Raton, FL, 1992.

Materials

▲   Cobalt(II) chloride hexahydrate, $CoCl_2 \cdot 6H_2O$), powder.

▲   Potassium hexacyanoferrate(III), $K_3[Fe(Cn)_6]$, powder.

▲   Copper(II) chloride dihydrate, $CuCl_2 \cdot 2H_2O$, powder.

▲   Silver nitrate, $AgNO_3$, powder.

▲   Nickel(II) nitrate, $Ni(NO_3)_2$, powder.

▲   Sodium hydroxide, NaOH, crystals.

▲   Sodium carbonate, Na_2CO_3, powder.

▲   Potassium carbonate, K_2CO_3, powder.

Uncommon Cobalt Complexes

When potassium thiocyanate is added to solutions of cobalt(II) ions, the familiar and beautiful bright blue complex is formed. The addition of water or nonpolar organic solvents changes the structure (and complex) and thus the resulting colors.

Procedure

Please consult the Safety Information before proceeding.

1. Pour all 15 mL of saturated cobalt chloride solution into a test tube.

2. Dropwise, swirling after each addition, add saturated potassium thiocyanate solution until a bright blue color is obtained. Then add 3 more drops.

3. Divide the blue solution equally into three test tubes. To one test tube, add distilled water dropwise, swirling after each addition until a violet color is obtained.

3. Cobalt salts by ingestion produce nausea and vomiting by local irritation. They are poisonous by subcutaneous, intravenous, and intraperitoneal routes. They are moderately toxic by ingestion. They are experimental teratogens; that is, reproductive effects have occurred.

4. Potassium thiocyanate may cause skin eruptions and psychosis.

5. Never use a solution of commercial glyoxal that is stronger than 40% glyoxal. Glyoxal can be absorbed through the skin and is also flammable. It is moderately irritating to the skin and mucous membranes. Work in an efficient fume hood when preparing solutions of this substance.

Materials

▲ Saturated cobalt(II) chloride hexahydrate, $CoCl_2 \cdot 6H_2O$: Dissolve 15 g (1/2 tsp) in 15 mL of water.

▲ Saturated potassium thiocyanate, KSCN: Add 15 g to 10 mL of water. Stir until as much as possible is dissolved.

▲ Ethanol, C_2H_5OH: 100% commercial.

▲ Acetone, $(CH_3)_2CO$: 100% commercial.

▲ Glyoxal, 40% CHOCHO: Order as 40%. A precipitate may form in this solution because of prolonged storage but can be redissolved by warming to 50–60 °C.

▲ Methanol, CH_3OH: 100% commercial.

4. In the second test tube, double the volume with water. In the third test tube, double the volume with any one of the following: acetone, methanol, ethanol, or 40% glyoxal. Compare the results.

Concepts

▲ The complexes formed with SCN^- ions are not clearly understood and certainly not accepted by all researchers. The amount of water that is complexed with the cobalt to give it a pink to violet color may be 6, 5, 4, or 3 molecules. It is fairly well accepted that the initial blue color is due to a square-planar arrangement of the four SCN^- ions in an aqueous solution.

▲ To get the results described under Procedures, the initial solutions of KSCN and $CoCl_2$ are saturated. As water is added to the square-planar SCN^- complex, the color reverts to the more familiar pink color of the octahedral complex. The colors can range from pink to violet, depending on the ratio of SCN^- to H_2O.

▲ When one of the organics is added to the original square-planar tetrathiocyanocobaltate(II) complex, $[Co(SCN)_4]^{2-}$, the structure changes to the tetrahedral shape and the blue becomes more intense.

▲ Although cobalt has been thoroughly studied, not all authors agree on the cause–effect relationship of structure and color.

Reactions

1. $Co^{2+}(aq) + 4SCN^-(aq) \rightarrow [Co(SCN)_4]^{2-}(aq)$
　　　pink　　　　colorless　　　　　　blue

2. $[Co(SCN)_4]^{2-}(aq) + 4H_2O \rightarrow$
　　　blue　　　　　$Co(SCN)_2(H_2O)_4(aq) + 2SCN^-(aq)$
　　　　　　　　　　　pink–violet

3. $Co(SCN)_4{}^{2-}(aq) + (excess)H_2O \rightarrow$
　　　blue　　　　　　　　　$Co(H_2O)_6{}^{2+}(aq) + 4SCN^-(aq)$
　　　　　　　　　　　　　　pink

4. $Co(SCN)_4{}^{2-}(aq) \xrightarrow[\text{solvent}]{\text{organic}} Co(SCN)_4{}^{2-}(solvated)$
　　　blue　　　　　　　　　　　　　　　　deep blue

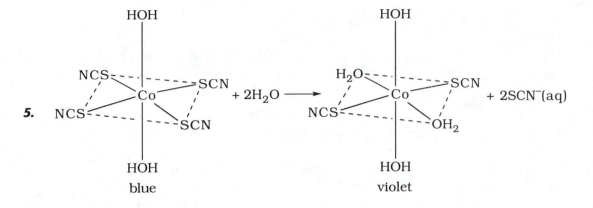

5. blue + 2H₂O ⟶ violet + 2SCN⁻(aq)

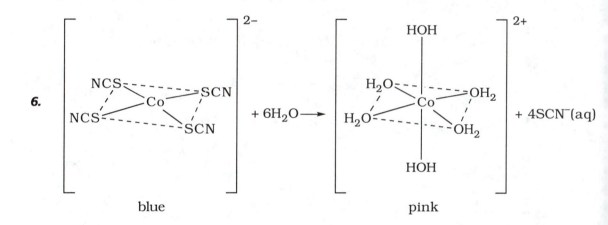

6. blue + 6H₂O ⟶ pink + 4SCN⁻(aq)

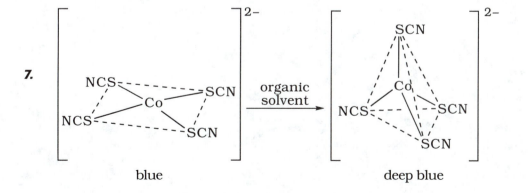

7. blue —organic solvent→ deep blue

Notes

The cobalt(II) chloride and potassium thiocyanate solutions must be saturated for the best effect.

References

Feigl, F. *Spot Tests, Vol. I, Inorganic Chemistry*, 4th English translation; Elsevier: Houston, TX, 1954.

Greenwood, N. N.; Earnshaw, A. *Chemistry of the Elements;* Pergamon: New York, 1984; pp 1311–1316.

Nassau, K. *The Physics and Chemistry of Color;* Wiley-Interscience: New York, 1983; p 308.

Ophardt, C. E. *J. Chem. Educ.* **1980**, *57*, 453.

Zeltman, A. H.; Matwiyoff, N. A.; Morgan, L. L. *J. Phys. Chem.* **1968**, *72*, 121.

Acidic Decomposition of Ultramarine

When a few drops of acid are added to ordinary laundry bluing, a rotten-egg odor is produced.

Procedure

Please consult the Safety Information before proceeding.

1. Place 3 mL of laundry bluing in a test tube and dilute it with 3 mL of distilled water. Shake the tube thoroughly. Notice that the contents of the test tube are in a suspension.

2. Add 5 drops of hydrochloric acid and mix the suspension thoroughly. In about 30–45 s, the suspension will begin to turn white and a precipitate will form.

3. **Carefully** detect the odor of the test tube by wafting.

Materials

▲   Hydrochloric acid, 0.1 M HCl: Carefully add 6.3 mL of concentrated acid to enough water to make 100 mL of solution.

▲   Bluing: commercial laundry bluing.

Reactions

$$Na_8S_2(Al_6Si_6O_{24}) + H^+(aq) \rightarrow$$
$$H_2S(g) + silicates(gel)$$

Concepts

▲   Commercial laundry bluing contains ultramarine as its active ingredient. Ultramarine is a framework silicate that has silicon and aluminum atoms at the corners of a polyhedron. Sodalite, the naturally occurring mineral, is white with the formula $Na_8(Cl_2)(Al_6Si_6O_{24})$, with chloride ions in place of sulfides that would be found in ultramarine. As the chloride ions are partially replaced with sulfide ions, the color becomes the brilliant blue of the mineral lapis lazuli.

▲   Bluing's ability to whiten is the result of color absorption and reflection. The wavelength of light reflected from dingy yellow clothes is white minus blue. The S_2^- and S_3^- in the ultramarine probably account for the blue color and absorb strongly in the yellow range and reflect blue. However, the blue is very faint and tends to show up as bluish white. The yellow wavelengths were absorbed and therefore cannot be reflected, and the laundry no longer appears to be dingy yellow. Modern whiteners absorb in the visible range and re-emit in the ultraviolet range and thereby make the clothing appear very white, almost glowing.

▲   Soaps and detergents are basic and do not affect the ultramarine. However, weak acids will quickly attack the molecule and release hydrogen sulfide gas.

▲   One recipe from the 1800s used to prepare ultramarine is 540 kg of kaolin, 8 kg of caustic soda, 538 kg of soda ash, 268 kg of sulfur, and 46 kg of charcoal.

Notes

At one time several different types of bluings were available: ultramarine, Prussian blue, and aniline dye. All of the compounds were suspended in an aqueous system and added to the clothes before washing and in the rinse water. We were able to find only one modern bluing agent with ultramarine. In antique stores, one might still be able to find the older varieties.

References

Greenwood, N. N.; Earnshaw, A. *Chemistry of the Elements;* Pergamon: New York, 1984; pp 406, 416.

Kingzett, C. T *Chemical Encyclopedia;* Van Nostrand: New York, 1928; p 748.

Partington, J. R. *A Textbook of Inorganic Chemistry;* Macmillan: London, 1950; p 813.

Ferrates by Electrolysis

When a saturated solution of potassium hydroxide is electrolyzed with an iron electrode, a pale red solution of potassium ferrate is produced.

Procedure

Please consult the Safety Information before proceeding.

1. Place 50 mL of a saturated solution of potassium hydroxide in a 100-mL tall-form beaker on an overhead projector.

2. Place an iron electrode and a carbon electrode in the beaker. Make the iron electrode the anode by connecting it with a patch cord to the positive terminal of a 100-V dc source. Make the carbon electrode the cathode by connecting it with a patch cord to the negative terminal of the 100-V dc source. Clamp the electrodes so that they cannot touch.

Safety Information

1. Potassium hydroxide is corrosive to all tissues. Inhalation of the dust or concentrated mist may cause damage to the respiratory tract. Wear gloves, an apron, and goggles when handling.

2. A voltage of 100 V dc is dangerous and must be handled with care. If the batteries are short-circuited, they will heat up enough to cause severe burns, and there is always the danger of an overheated battery exploding. The 100-V potential difference can also cause damage to the skin and may severely affect the heart.

Materials

▲ Potassium hydroxide, saturated KOH: Carefully dissolve 30 g in enough water to make 100 mL of solution.

▲ Electrolysis apparatus.

3. Observe the deep red color rising to the surface with the bubbles. If the color is not readily apparent, reverse the patch cords for 1 min. As the tiny stream of color rises, it spreads out and makes a very light red (pink) solution.

Concepts

▲ Anodic oxidation at the iron electrode produces iron(VI). Cast iron works well because it is covered with black iron oxide, $FeO \cdot Fe_2O_3$. This helps to prevent a side reaction with the concentrated hydroxide. With the 100-V potential difference, anodic oxidation takes place, producing the ferrate ion on or near the surface of the iron electrode.

▲ Cathodic reduction can be observed at the site of the carbon electrode, where a copious amount of hydrogen gas is produced.

▲ The solution quickly loses its pale red color because the ferrate ion is unstable.

Reactions

1. $Fe(s) + 2OH^-(aq) + 2H_2O(l) \xrightarrow{\text{electrolysis}} (FeO_4)^{2-}(aq) + 3H_2(g)$

2. $2H_2O(l) \xrightarrow{\text{electrolysis}} 2H_2(g) + O_2(g)$

3. $4(FeO_4)^{2-}(aq) + 6H_2O(l) \rightarrow 2Fe_2O_3 \cdot H_2O(s) + 8OH^-(aq) + 3O_2(g)$

Notes

1. Sometimes the color does not appear quickly; an unexplained remedy is to reverse the patch cords for about 1 min.

2. A safe and simple method for obtaining a 100-V dc source is as follows. Snap together male and female connectors on eleven 9-V alkaline batteries. Place tape over the open connector on the first battery until the series is completed.

This precaution will help prevent a short circuit. A desk-top power source may not supply enough amperage for this demonstration.

References

Greenwood, N. N.; Earnshaw, A. *Chemistry of the Elements;* Pergamon, New York, 1984; pp 1254, 1256–1257, 1265.

Partington, J. R. *A Textbook of Inorganic Chemistry;* Macmillan: London, 1950; p 935.

Nickel in the 4+ State

When a few drops of colorless aqueous ammonia and dimethylglyoxime are added to a bright grass-green solution, a familiar scarlet precipitate forms. If, instead, a reddish solution of bromine water is added to the bright grass-green solution followed by aqueous ammonia, a deep orange–red solution appears.

Procedure

Please consult the Safety Information before proceeding.

1. Add 5 mL of nickel chloride solution to two test tubes. To one test tube add 1 drop of aqueous ammonia and 5 drops of dimethylglyoxime. Observe the formation of the classic scarlet precipitate.

2. To the second test tube add 5 mL of bromine water. Dropwise, add aqueous ammonia until

3. Concentrated ammonia, when inhaled, will cause edema of the respiratory tract, spasm of the glottis, and asphyxia. Treatment must be prompt to prevent death.

4. Nickel salts are toxic by ingestion. They are poisonous by subcutaneous, intravenous, and intraperitoneal routes. They are experimental teratogens and shows experimental reproductive effects.

the color of the bromine has disappeared. Then add 5 drops of dimethylglyoxime and shake. Observe the orange–red solution that forms instead of the scarlet precipitate.

Concepts

▲ Dimethylglyoxime (DMG) is a chelating compound, which is very suitable for attaching metal ions. The structure appears to have claws like a lobster's where the nitrogens are located. The nitrogens are electron deficient and trap the nickel ion by using nickel's d electron pairs. The alkaline DMG produces a scarlet insoluble complex. This procedure is the standard qualitative method for identifying nickel ions.

▲ When bromine water is added to the nickel solution, it oxidizes the nickel to the 4+ state and chelates with DMG, but it also bonds with an oxygen. This compound is orange–red in color and is soluble. Greenwood and Earnshaw (1984) mentioned that the nickel is probably best described as being nickel(II) with an oxidized ligand, although the complex formally contains nickel(IV).

▲ Structures of the scarlet precipitate and orange–red solution are

scarlet precipitate   orange–red solution

Reactions

1. $Ni^{2+}(aq) + DMG(C_2H_5OH) + OH^-(aq) \rightarrow Ni^{II}DMG(s)$
                                                        scarlet

2. $Ni^{2+}(aq) + Br_2(aq) \rightarrow Ni^{4+}(aq) + 2Br^-(aq)$

3. $Ni^{4+}(aq) + DMG(C_2H_5OH) \rightarrow Ni^{IV}DMG(aq)$
                                        orange-red

4. $Br_2(aq) + OH^-(aq) \rightarrow HOBr(aq) + Br^-(aq)$

Notes

This demonstration is excellent to introduce ligands, chelating agents, and qualitative analysis theory.

References

Feigl, F. *Spot Tests, Vol. I, Inorganic Chemistry*, 4th English translation; Elsevier: Houston, TX, 1954; pp 227, 296.

Greenwood, N. N.; Earnshaw, A. *Chemistry of the Elements;* Pergamon: New York, 1984; pp 1339–1340.

Partington, J. R. *A Textbook of Inorganic Chemistry;* Macmillan: London, 1950; p 949.

Materials

▲ Nickel chloride, 0.1 M $NiCl_2$: Dissolve 2.4 g in enough water to make 100 mL of solution.

▲ Dimethylglyoxime (DMG), 1.0% $C_4H_8N_2O_2$: Dissolve 1.0 g in enough ethanol to make 100 mL of solution.

▲ Aqueous ammonia, 1 M $NH_3(aq)$: Dissolve 6.7 mL of concentrated ammonia in enough water to make 100 mL of solution.

▲ Bromine water, saturated: Place 1 g of pyridinium bromide perbromide in 10 mL of water in a test tube. Wait about 10 min for the solution to become saturated.

Ferrates by Fusion

When iron powder and solid potassium nitrate are fused and the resulting product is placed in water, a purple solution can be extracted.

Procedure

Please consult the Safety Information before proceeding.

1. Place about 5 g (1/2 tsp) of powdered potassium nitrate in a crucible. Add about 3 g (1/3 tsp) of iron powder and mix the contents thoroughly by shaking or stirring with a wooden splint. Place the resulting mixture in a Pyrex test tube.

2. Heat the mixture carefully with a Bunsen burner, behind a safety shield, until the mixture ignites. In a few seconds, when the exothermic reaction is complete, allow the tube to cool to room temperature.

Materials

▲ Potassium nitrate, powder.

▲ Barium nitrate, 0.1 M $Ba(NO_3)_2$: Dissolve 0.3 g in enough water to make 10 mL of solution.

▲ Iron powder.

3. Add 15 mL of distilled water. Stir the mixture to dissolve as much of the residue as possible. Decant the deep purple solution as quickly as possible into two separate test tubes.

4. Add several drops of barium nitrate solution to one test tube and observe the red precipitate. Observe that the precipitate remains red, whereas the other solution quickly becomes reddish brown.

Concepts

▲ Iron(VI) is an uncommon oxidation state for iron and can be made only with some difficulty by fusing iron with potassium nitrate, a strong oxidant. The exothermic reaction, once initiated, will produce a glowing mass in the bottom of the test tube.

▲ When barium nitrate is added to the deep purple solution containing ferrate, FeO_4^{2-} ion, a red precipitate of $BaFeO_4$ forms.

▲ Ferrate ions are unstable, and Fe(VI) is reduced at pH above 2–3. Fe(III) ions react with OH⁻ ions to produce a colloidal gel of $Fe(OH)_3$, which will slowly form a reddish-brown precipitate of Fe_2O_3. The FeO_4^{2-} decomposes into Fe_2O_3 and O_2 gas in the test tube, as evidenced by the formation of the precipitate without the visible intermediate formation of the colloidal gel.

Reactions

1. $Fe(s) + 2KNO_3(s) \rightarrow K_2FeO_4(s) + 2NO(g)$
2. $K_2FeO_4(s) + H_2O(l) \rightarrow 2K^+(aq) + FeO_4^{2-}(aq)$
3. $4FeO_4^{2-}(aq) + 6H_2O(l) \rightarrow$
      $2Fe_2O_3 \cdot H_2O(s) + 8OH^-(aq) + 3O_2(g)$
4. $Ba^{2+}(aq) + FeO_4^{2-}(aq) \rightarrow BaFeO_4(s)$

Notes

1. The classic method of synthesis of a ferrate is by the oxidation of a concentrated alkaline sus-

pension of freshly prepared iron(III) oxide hydrate with chlorine gas. Although this method is possible in the high-school laboratory, the production of sufficient quantities of chlorine gas is not easily or safely accomplished.

2. The most easily made iron(VI) compound is called a "ferrite", which is a mixture of metal oxides, having the normal spinel structure common to garnets and having substantial technological importance. The common magnetic iron oxide, $FeO \cdot Fe_2O_3$, a mixture of two iron(III) ions to one iron(II) ion, has the reverse spinel structure.

References

Greenwood, N. N.; Earnshaw, A. *Chemistry of the Elements;* Pergamon: New York, 1984; p 278–281, 1254, 1256–1257, 1265.

Partington, J. R. *A Textbook of Inorganic Chemistry;* Macmillan: London, 1950; p 935.

Cerium Hydroperoxide

When hydrogen peroxide is added to an alkaline solution of cerium ions, an intensely orange precipitate is formed.

Procedure

Please consult the Safety Information before proceeding.

1. Add 5 mL of a solution of a cerium salt to a test tube. Then add several drops of aqueous ammonia. Shake the solution.

2. Dropwise add 9% hydrogen peroxide until an orange precipitate forms. Then add 2 more drops. Shake the solution thoroughly.

Concepts

▲ The hydrogen peroxide produces a small amount of hydroperoxide ion, OOH^-, by

Materials

▲ Ammonium hexanitrato-cerate(IV), 0.1 M $Ce(NH_4)_2$-$(NO_3)_6$: Dissolve 5.4 g in enough water to make 100 mL of solution.

▲ Aqueous ammonia, 1.0 M NH_3: Add 6.7 mL of concentrated stock aqueous ammonia to enough water to make 100 mL of solution.

▲ Hydrogen peroxide, 9%, H_2O_2: Add 3 mL of 30% hydrogen peroxide to enough water to make 10 mL of solution or purchase 9% hydrogen peroxide (Clairoxide) from a hair stylist.

deprotonation in basic solutions. This small amount reacts with cerium hydroxide to produce the insoluble hydroperoxide. Because the product formed is insoluble, the dissociation of hydrogen peroxide shifts to the right and the cerium–hydroperoxide reaction goes to completion.

▲ Oxygen bridges are common for some transition elements, in particular, lanthanum and cobalt, with large organic complexes. Possibly the cerium forms oxygen bridges between adjacent cerium ions to produce the polymeric gelatinous $CeO_2 \cdot xH_2O$.

Reactions

1. $Ce^{4+}(aq) + 3OH^- + H_2O_2(aq) \rightarrow$
$$Ce(OH)_3(OOH)(s) + H^+(aq)$$
orange

2. $2Ce^{3+}(aq) \rightarrow Ce^{2+}(aq) + Ce^{4+}(aq)$

Notes

1.   This demonstration can be used to show an unusual complex of a hydroperoxide compound and the autoxidation of a cerium salt.

2.   If cerium(IV) is unavailable, cerium(III) can be used. Cerium(III) hydroxide undergoes disproportionation to cerium(II) and cerium(IV). This solution then can be reacted with hydrogen peroxide.

References

Feigl, F. *Spot Tests, Vol. I, Inorganic Chemistry*, 4th English translation; Elsevier: Houston, TX, 1954; pp 197–198.

Partington, J. R. *A Textbook of Inorganic Chemistry*; Macmillan: London, 1950; pp 821–822.

Greenwood, N. N.; Earnshaw, A. *Chemistry of the Elements*; Pergamon: New York, 1984; pp 719, 1444–1445.

Colored Compounds of Metal and Diphenylthiocarbazone

When diphenylthiocarbazone is added to separate solutions of metal ions, insoluble complex ions are produced that are soluble in the organic solvent trichlorotrifluoroethane (TTE).

Procedure

Please consult the Safety Information before proceeding.

1. Place 3 mL of TTE in each of three test tubes. Add 3 drops of diphenylthiocarbazone indicator to each tube and shake the tube. The solutions will be deep green.

2. Add 1 mL of zinc solution to one test tube (raspberry red); 1 mL of silver solution to the second test tube (lavender), and 1 mL of copper solution to the third test tube (yellow–brown). Stopper and shake the tubes thoroughly until a brightly colored TTE layer forms.

Safety Information

1. Copper sulfate is a strong irritant and toxic if ingested. It may be harmful if it comes in contact with mucous membranes or if the dust is inhaled.

2. Ethanol is a volatile, highly flammable liquid. It should not be used near any open flames. Prolonged topical use may cause dryness, and inhalation may produce headaches. →

3. Silver nitrate is corrosive and will stain the skin. Handle with care by using gloves and an apron, avoiding all contact with the skin. Inhalation of the dust results in corrosive action on the mucous membranes.

4. Carbon tetrachloride is toxic; use only in a well-ventilated hood. TTE can be used in a well-ventilated room; however, avoid prolonged use.

Concepts

▲ The diphenylthiocarbazone can form a complex with several metal ions by bonding with the metal ion with one nitrogen and chelating with the other nitrogen. With copper and zinc, the metal ion is complexed with two diphenylthiocarbazones. With silver ions, the complex is formed with one diphenylthiocarbazone.

▲ The interesting aspect of these complexes is that they form so readily and show different solubilities and colors in water and organic solvents.

Ion	Water	Carbon Tetrachloride	TTE
Zn^{2+}	colorless	raspberry red	raspberry red
Ag^+	colorless	red–orange	lavender
Cu^{2+}	blue	yellow	yellow–brown

▲ Diphenylthiocarbazone can be considered to be showing characteristics of an indicator, that is, breaking double bonds, changing the location of unbonded electrons, and changing the number or location of hydrogens. Any one of these behaviors can change the color of the indicator. The diphenylthiocarbazone actually complexes with the metal.

▲ CCl_4 and TTE are nonpolar organic solvents; thus, the indicator and bonded metal are more soluble in the organic solvent than in the very polar water.

Reactions

1. $Zn^{2+}(aq) + 2C_{13}H_{12}N_4S \rightarrow$
$$Zn(C_{13}H_{11}N_4S)_2 + 2H^+(aq)$$

2. $Cu^{2+}(aq) + 2C_{13}H_{12}N_4S \rightarrow$
$$Cu(C_{13}H_{11}N_4S)_2 + 2H^+(aq)$$

3. $Ag^+(aq) + C_{13}H_{12}N_4S \rightarrow Ag(C_{13}H_{11}N_4S) + H^+(aq)$

4.

diphenylthiocarbazone (green)   diphenylthiocarbazonate (raspberry red)

Notes

1. Zinc, copper, and silver ions are most likely to be available to the high-school teacher, but lead and mercury give excellent results as well.

2. Carbon tetrachloride may not be available to many demonstrators, but the colors are given here in case you have access to carbon tetrachloride.

References

Feigl, F. *Spot Tests, Vol. I, Inorganic Chemistry*, 4th English translation; Elsevier: Houston, TX, 1954; pp 273, 327.

Nassau, K. *The Physics and Chemistry of Color;* Wiley: New York, 1983.

Tomicek, O. *Chemical Indicators;* Butterworths: London, 1951.

Materials

▲ Diphenylthiocarbazone (dithizone) indicator, 0.1% $C_{13}H_{12}N_4S$: Dissolve 0.1 g in enough ethyl alcohol to make 100 mL of solution.

▲ Zinc nitrate hexahydrate, 0.5 M $Zn(NO_3)_2 \cdot 6H_2O$: Dissolve 15 g in enough water to make 100 mL of solution.

▲ Silver nitrate, 0.5 M $AgNO_3$: Dissolve 7.5 g in enough water to make 100 mL of solution.

▲ Copper(II) nitrate trihydrate, 0.5 M $Cu(NO_3)_2 \cdot 3H_2O$: Dissolve 12 g in enough water to make 100 mL of solution.

▲ Carbon tetrachloride, 100% CCl_4: commercial solution.

▲ 1,1,1-Trichloro-2,2,2-trifluoroethane, $C_2Cl_3F_3$: commercial solution.

Kinetics and Equilibrium

Glyoxal Clock Reaction

In the presence of an indicator, mixing 40% glyoxal with a metabisulfite–sulfite solution results in a color change that can be timed. Several color changes occur before the final color is reached.

Procedure

Please consult the Safety Information before proceeding.

1. Measure 400 mL of distilled water into a 500-mL beaker.

2. Place 1 mL of indicator in the beaker and stir the solution to mix thoroughly.

3. Place 5 mL of metabisulfite–sulfite solution in the beaker and stir the solution to mix.

4. On the addition of the next solution, begin timing. Place 5 mL of glyoxal solution in the beaker and stir the solution to mix.

5. Note the time when the color change is complete.

Safety Information

1. Glyoxal solution, 40%, is moderately irritating to the skin and mucous membranes. Work in an efficient fume hood when preparing solutions of this substance.

2. Sodium metabisulfite, sodium sulfite, and EDTA may by harmful by inhalation, ingestion, or skin absorption.

3. Universal indicator is toxic by inhalation.

Materials

Note: To give repeatable results throughout 1 day, the solutions should be prepared 24 h in advance.

▲ Sodium metabisulfite-sodium sulfite, $Na_2S_2O_5$-Na_2SO_3: In a 250-mL volumetric flask, dissolve 4.5 g of $Na_2S_2O_5$, 0.9 g of disodium ethylenediaminetetraacetate (Na_2EDTA), and 0.75 g of anhydrous Na_2SO_3 in enough distilled water to make 250 mL of solution.

▲ Glyoxal, CHOCHO: In a 250-mL volumetric flask, add 23 mL of 40% glyoxal to enough distilled water to make 250 mL of solution. Prepare this solution in a fume hood. A precipitate may form in the lab bottle of glyoxal because of prolonged storage but can be redissolved by warming to 50–60 °C.

▲ Indicators: phenol red, universal, bromthymol blue, neutral red, *m*-nitrophenol, cresol red.

Concepts

▲ In the past, this reaction was done with formalin, a 40% formaldehyde solution, instead of glyoxal and was called the "formaldehyde clock". Formaldehyde is a known carcinogen, and on the basis of a suggestion by the late Miles Pickering, we tried 40% glyoxal as a substitute. The similarity in the two substances can be seen from the structural formulas:

formaldehyde   glyoxal

▲ The reaction goes from slightly acidic to slightly basic, about pH 5.5 to 8.1. The choice of indicators is restricted to the pH range of the reaction. Following are indicators that can be used:

Indicator	Color in Acid	Color in Base	pH Range
Phenol red	yellow	pink	6.8–8.4
Universal	yellow–orange	lime-green	1.0–14.0
Bromthymol blue	yellow	blue	6.0–7.6
Neutral red	red	amber	6.8–8.0
m-Nitrophenol	colorless	yellow	6.8–8.6
Cresol red	yellow	red	7.0–8.8

▲ In a "clock" reaction, the time it takes for the reaction to occur can be determined because a color change indicates the reaction is complete—hence the use of an indicator. The rate of the reaction, or the kinetics, can be determined by varying the concentration of the reactants. When the concentration of only one reactant is varied at a time, while the others are kept constant, the effect of each reactant on the rate of the reaction can be determined.

▲ The metabisulfite–sulfite solution is readily oxidized in air. The Na_2EDTA helps preserve the solution's concentration and gives repeatable results up to, but not past, 24 h.

Notes

1. Never use a solution of commercial glyoxal that is stronger than 40%. Glyoxal can be absorbed through the skin and is also flammable.

2. You can determine the effect of changing temperature by varying the temperature between 4 and 38 °C. Using 5 mL of glyoxal, 5 mL of metabisulfite–sulfite, and 414 mL of water, the temperatures we used and resulting times were as follows:

Temperature (°C)	Time (s)
4	78
9	48
14	30
19	20
24	13
28	10
33	7
38	5

3. You can determine the effect of changing concentration by keeping the volume of one solution constant and varying the volume of the other solution. Using 1 mL of indicator, following are the volumes of reactants we used and average time we determined using a stopwatch and verified with a pH probe interfaced with a computer:

Metabisulfite–Sulfite	Glyoxal (mL)	Water (mL)	Average Time (s)
5.0	1.5	417.5	65
5.0	2.5	416.5	51
5.0	5.0	414.0	22
5.0	10.0	409.0	17
5.0	15.0	404.0	11
1.5	5.0	417.5	17
2.5	5.0	416.5	20
5.0	5.0	414.0	22
10.0	5.0	409.0	39
15.0	5.0	404.0	62
20.0	5.0	399.0	90
25.0	5.0	394.0	135

4. The solutions give repeatable results 24 h after preparation but do not work after 48 h.

5. Arrange enough beakers in a row, based on the number of different indicators you have, so students can observe the various color changes.

References

Burnett, M. G. *J. Chem. Educ.* **1982**, *59*, 160.

Cassen, T. *J. Chem. Educ.* **1976**, *53*, 197.

Ealy, Julie B. *Sci. Teach.* **1991**, *58*(4), 26.

Ealy, Julie B.; Ealy, James L., Jr. *Close-Up on Chemistry*, ACS video manuscript; American Chemical Society: Washington, DC, 1991.

Solutions That Glow in the Dark

After fluorescent dyes are dissolved in solutions of luminol–dimethyl sulfoxide, addition of a colorless solution of sodium hydroxide with shaking causes the resulting solutions to glow in the dark.

Procedure

Please consult the Safety Information before proceeding.

1. In a test tube, dissolve 0.003 g of dye in 10 mL of luminol–dimethyl sulfoxide solution.

2. With a microtip pipet, add 12 drops of sodium hydroxide to the test tube.

3. Stopper the test tube and shake it vigorously to mix the two solutions.

4. Using two more test tubes, repeat Steps 1–3 with the other two dyes.

5. Turn out the lights and display the glowing red, yellow, and green test tubes.

Safety Information

1. Dimethyl sulfoxide is readily absorbed through the skin and results in primary irritation with redness, itching, and sometimes scaling. Wear rubber gloves. It can transport materials through the skin. Contact with the skin or eyes must be avoided.

2. Sodium hydroxide is corrosive to all tissues. Inhalation of the dust or concentrated mist may cause damage to the respiratory tract. Wear gloves, goggles, and a rubber apron when handling.

3. Toxic and carcinogenic properties of luminol, 5,12-bis(phenylethynyl)naphthacene, and 9,10-bis(phenylethynyl)anthracene are not known. →

4. Fluorescein is toxic to all tissues. Inhalation of the dust or concentrated mist may cause damage to the respiratory tract. Wear gloves, a rubber apron, and goggles when handling.

6. When the glowing diminishes, open the test tube to introduce oxygen, stopper the tube, and shake the tube again.

Concepts

▲ Structural formulas of the fluorescent compounds are

fluorescein
(yellow)

5,12-bis(phenylethynyl)naphthacene
(red)

9,10-bis(phenylethynyl)anthracene
(green)

Materials

▲ Luminol (3-amino-phthalic hydrazide), powder.

▲ Luminol–dimethyl sulfoxide, $C_8H_7O_2N_3$–C_2H_6OS: Dissolve 0.5 g of luminol in 100 mL of dimethyl sulfoxide.

▲ Sodium hydroxide, saturated NaOH: Dissolve 50 g of NaOH in 50 mL of distilled water. Cool to room temperature before using.

▲ Fluorescein, powder.

▲ 5,12-Bis(phenylethynyl)naphthacene, powder.

▲ 9,10-Bis(phenylethynyl)anthracene, powder.

▲ When dimethyl sulfoxide is used, chemiluminescence results from luminol reacting with molecular oxygen dissolved in the dimethyl sulfoxide.

▲ In basic solution, the bonds between the nitrogen and hydrogen atoms of the two –NH groups on the luminol molecule are broken and produce the dianion of luminol. The hydrogens combine with OH⁻ ions to form water.

▲ Dimethyl sulfoxide causes the luminol dianion to form the excited-state product, and N_2 gas is released. The excited-state product decays to the ground state, emitting energy in the form of light.

▲ Because the emission of visible light of a specific wavelength requires a specific amount of energy, a chemical reaction cannot produce light unless it is highly energy-releasing. A species must receive the excitation energy produced by the reaction, resulting in an electronically excited state. The species that is produced as an intermediate can pass on its excess energy to excite a fluorescer, which then emits visible light.

▲ In 1603, Vincenzo Cascariolo heated a mixture of barium sulfate and coal. After being cooled, the powder exhibited a bluish glow at night. The glow was restored by exposure of the powder to sunlight. Lapis solaris was the name given to the powder.

Reactions

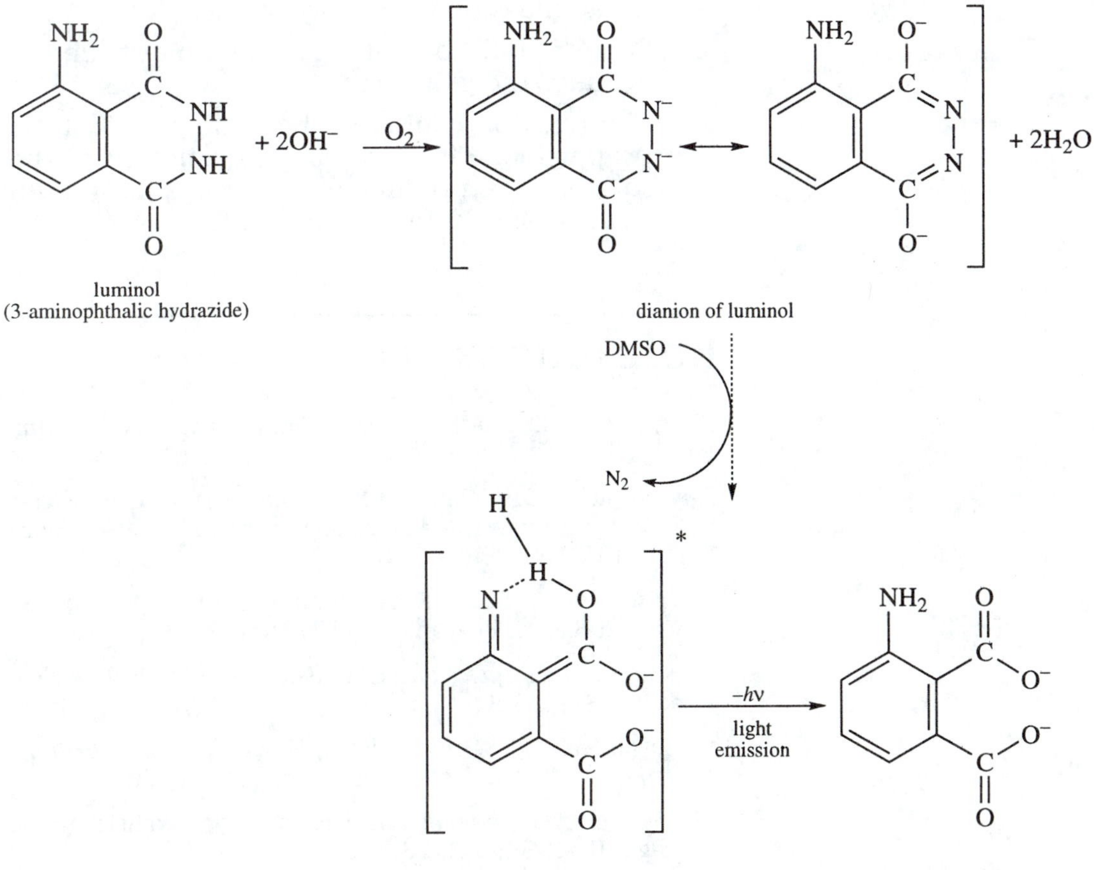

luminol
(3-aminophthalic hydrazide)

dianion of luminol

$+ 2H_2O$

DMSO

N_2

excited state of
3-aminophthalate ion

ground state of
3-aminophthalate ion

Notes

1. The luminol–dimethyl sulfoxide solution can be kept at least 2 months at room temperature. Although it darkens somewhat, it still reacts very well.

2. Other fluorescent compounds can be used with the following color results: rubrene, salmon; rhodamine B, rose; eosin Y, orange; and luminol alone, bluish.

3. You might consider choosing fluorescent compounds that represent your school colors or the colors of holiday seasons.

4. The test tubes can be stoppered and kept refrigerated for several days. The dimethyl sulfoxide will freeze. To activate the chemiluminescence, thaw the solutions, aerate them by removing the stopper, stopper the test tube, and shake the tube vigorously.

5. Oxygen could be bubbled into the test tube to produce the glowing.

6. The reaction could also be done in a flask by proportionately increasing the amounts of the solutions. Approximately 20 drops of sodium hydroxide solution equals 1 mL.

References

Chalmers, J. H.; Bradbury, M. C.; Fabricant, J. D. *J. Chem. Educ.* **1987**, *64*, 969.

Ealy, Julie B.; Ealy, James L., Jr. *Close-Up on Chemistry;* ACS video manuscript; American Chemical Society: Washington, DC, 1991; pp 19-20.

Morrison, R. T.; Boyd, R. N. *Organic Chemistry*, 3rd ed.; Allyn and Bacon: Boston, MA, 1973; pp 31–32.

Nassau, K. *The Physics and Chemistry of Color;* Wiley: New York, 1983; pp 360–361.

Nebergall, W. H.; Schmidt, F. C.; Holtzclaw, H. F. *College Chemistry*, 5th ed.; Heath: Lexington, MA, 1976; p 308.

The New Encyclopedia Britannica; Encyclopedia Britannica: Chicago, IL, 1989; Vol. 23, p 22.

Rosewell, D. F.; White, E. H. *Meth. Enzym.* **1978**, *57*, 409.

Schneider, H. W. *J. Chem. Educ.* **1941**, *18*, 347.

Schneider, H. W. *J. Chem. Educ.* **1970**, *47*, 519-522.

Shakhashiri, B. Z. *Chemical Demonstrations: A Handbook for Teachers of Chemistry;* University of Wisconsin: Madison, WI, 1983; Vol. 1, p 128.

Colorless, Hot and Blue, Cold Starch–Iodine Complex

Solutions of starch and iodine are mixed in a test tube, and a blue–black color results. When the solution is heated, the blue–black color disappears, but it returns when the test tube is placed in an ice-cold water bath.

Procedure

Please consult the Safety Information before proceeding.

1. Place 20 mL of starch solution in a test tube. Add 1.5 mL of iodine solution to obtain a blue–black color. Stir the solution.

2. Place the test tube in a 90 °C hot water bath until the color disappears or has a very pale yellow tinge.

3. Place the test tube in an ice-cold water bath. The dark blue color returns.

4. The process can be repeated a number of times.

Safety Information

1. Solid iodine is intensely irritating to the eyes, skin, and mucous membranes. The vapor causes edema of the respiratory tract, spasm of the glottis, and asphyxia.

2. Although starch is a food substance, no substance should ever be tasted in the laboratory.

3. Potassium iodide is moderately toxic by ingestion and intraperitoneal routes.

Materials

▲ Starch, $(C_6H_{10}O_5)_x$: Dissolve 1 g of starch in a few milliliters of water to make a smooth paste. Boil 200 mL of water and slowly, with stirring, add it to the starch. Cool before using.

▲ Iodine–potassium iodide solution, I_2–KI: Dissolve 1.3 g of I_2 and 4.0 g of KI in 50 mL of water. Dilute to 1000 mL.

Concepts

▲ The starch–iodine complex is amylose–iodine. Amylose is the linear starch fraction that is composed of chains of 1,4-linked α–D (+) glucopyranose units:

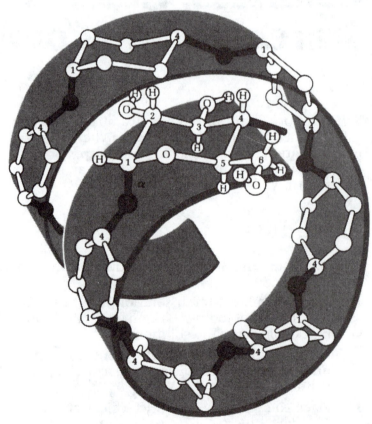

▲ The glucopyranose units coil back on each other, creating a loosely overlapping spiral with a central cavity or tube:

Illustration by Irving Geis. Reprinted with permission from Kemp, D. S.; Vellaccio, F. Organic Chemistry; Worth: New York, 1980; pp 993.

▲   The color of the complex, blue–black, comes
from the pentaiodide anion, I_5^-. Although normally an unstable anion, it becomes stable as a
part of the complex. It would be represented
as:

$$[I - I - - - I - - - I - I]^-$$

▲   Iodine molecules fit within the cavity of the
amylose helix:

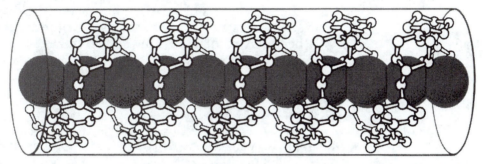

Illustration by Irving Geis. Reprinted with permission from Kemp, D. S.; Vellaccio, F.
Organic Chemistry; *Worth: New York, 1980; pp 994.*

▲   In this reaction, the addition of energy caused
by placing the test tube in a hot water bath shifts
the equilibrium in the forward direction or to
the right. This shift results in a color change
from blue to colorless. Removal of energy,
caused by placing the test tube in an ice bath,
shifts the equilibrium in the reverse direction,
or to the left, and causes the blue color to return.

Reactions

Hydrogen bonds exist between the amylose molecules. An increase in temperature disrupts the
bonds and destabilizes the helical structure. The
helical segments shorten, and the helix-to-nonhelix
ratio decreases. The amount of bound iodine decreases and the blue–black color therefore disappears as the temperature increases.

Suggested reaction:

starch–I_5^- complex + energy
        blue–black

$$I_2 \text{ (aq)} + \text{starch(aq)} + I_3^- \text{(aq)}$$
                        colorless

Notes

1. Place several drops of iodine solution at different locations on a piece of paper so students can observe the blue–black color that results from the presence of starch.

2. Have students observe that the colorless solution turns blue from the bottom to the top in the cold water and the blue–black starch–iodine complex swirls upward. In hot water, the opposite observations are noted as the blue solution becomes colorless from the top down.

References

Kemp, D. S.; Vellaccio, F. *Organic Chemistry;* Worth: New York, 1980; pp 993–994.

Newth, G. S. *Chemical Lecture Experiments;* Longmans, Green, & Co.: New York, 1892; p 107.

Radley, J. A. *Starch and Its Derivatives*, 4th ed.; Chapman & Hall: London, 1968; p 212.

Teitelbaum, R. C.; Ruby, S. L.; Marks, T. J. *J. Am Chem. Soc.* **1980**, *102*(10), 3322.

Bleach Plus One or Two Catalysts

Adding a blue copper(II) sulfate solution, or a green iron(II) sulfate solution, or both the blue and the green solutions to commercial bleach results in the evolution of gas.

Procedure

Please consult the Safety Information before proceeding. Perform this demonstration in a fume hood.

1. Using a hot water bath, heat 20 mL of liquid bleach in a test tube to 50–60 °C.
2. Add 1.0 mL of copper(II) sulfate solution.
3. Remove the tube from the hot water bath, stir to mix the solutions, and set the tube aside to make observations.
4. Repeat Steps 1–3 using 2.0 mL of iron(II) sulfate solution instead of copper(II) sulfate solution.

Safety Information

1. Copper sulfate penta-hydrate is toxic if ingested. It may be harmful if it comes in contact with mucous membranes or if the dust is inhaled.

2. Iron(II) sulfate hepta-hydrate may be harmful by inhalation, ingestion, or skin absorption. It causes eye and skin irritation. It is irritating to mucous membranes and the upper respiratory tract. →

3. Bleach or sodium hypochlorite solution is a corrosive liquid. It causes skin burns, and chlorine is evolved when bleach is heated. Inhalation may cause severe bronchial irritation and pulmonary edema. Prolonged skin contact may result in irritation. It is toxic by ingestion.

5. Compare the evolution of gas in the two test tubes.

6. Repeat Steps 1–3 using 1 mL of copper(II) sulfate solution and 1 mL of iron(II) sulfate solution.

7. Place the three test tubes in a test-tube rack to make observations.

Concepts

▲ Liquid bleach is an alkaline solution of NaOCl with a pH of at least 11, with about 5% available chlorine by weight.

▲ Hypochlorite anion, OCl⁻, with Cu^{2+} or Fe^{2+} acting as a catalyst, forms oxygen gas and chloride ions. More gas is evolved, though, when both the Cu^{2+} and Fe^{2+} solutions are combined with the bleach. Both of the catalysts together have a greater effect on the reaction than either catalyst alone.

▲ Secondary reactions are involved:

a. Copper(II) plus hydroxide ions form copper(II) hydroxide, and the heat of the overall reaction results in the formation of black copper(II) oxide.

b. Iron(II) plus hydroxide ions form iron(II) hydroxide. Oxygen produced in the reaction causes Fe^{2+} to be oxidized to Fe^{3+}, and amber-colored iron(III) hydroxide forms.

▲ Evidence exists for the chemical bleaching of cloth prior to 300 B.C. Commercial bleaching was begun by Bertholet in the 1790s after the discovery of chlorine by Scheele.

Reactions

1. $2OCl^-(aq) \xrightleftharpoons{Fe^{2+}/Cu^{2+}} O_2(g) + 2Cl^-(aq)$

2. $Cu^{2+}(aq) + 2OH^-(aq) \rightarrow Cu(OH)_2(s)$
        blue–green

3. $Cu(OH)_2(s) \rightarrow CuO(s) + H_2O(l)$
        black

This demonstration is attributed by Fowles to I. G. F. Joubert (1905).

4. $Fe^{2+}(aq) + 2OH^-(aq) \rightarrow Fe(OH)_2(s)$

5. $4Fe(OH)_2(s) + O_2(g) + 2H_2O(g) \rightarrow 4Fe(OH)_3(s)$

$$\text{amber}$$

Materials

▲ Copper sulfate pentahydrate, 0.1 M $CuSO_4 \cdot 5H_2O$: Dissolve 2.5 g in enough water to make 100 mL of solution.

▲ Iron(II) sulfate heptahydrate, 0.1 M $FeSO_4 \cdot 7H_2O$: Dissolve 2.0 g in enough water to make 100 mL of solution.

▲ Liquid bleach: From the supermarket; use as is, from the bottle.

Notes

1. Domestic bleach is a NaOCl solution that is 5% by weight. Although chlorine gas might be emitted when bleach is heated, using the water bath and keeping the temperature within the range suggested should avoid this problem. We could detect no odor of chlorine when performing this demonstration. Because the possibility does exist, though, the demonstration should be performed in a fume hood.

2. The emission of oxygen gas can be observed for several days. The convection currents initially set up in the test tubes are fascinating to watch.

References

Baum, B. M.; Finley, J. H.; Blumbergs, J. H.; Elliott, E. J.; Scholer, F.; Wooten, H. L. In *Kirk–Othmer Encyclopedia of Chemical Technology;* Wiley: New York, 1978; Vol. 3, p 938.

Fowles, G. *Lecture Experiments in Chemistry;* Blakiston: Philadelphia, PA, 1937; pp 469–470.

Greenwood, N. N.; Earnshaw, A. *Chemistry of the Elements;* Pergamon: New York, 1984; pp 1003–1007.

Nebergall, W. H.; Schmidt, F. C.; Holtzclaw, H. F., Jr. *College Chemistry*, 5th ed.; Heath: Lexington, MA, 1976; p 907.

A Milky Clock

Adding a constant volume of hydrochloric acid to six different beakers containing various amounts of sodium thiosulfate solution results in the formation of a milky solid in varying amounts of time.

Procedure

Please consult the Safety Information before proceeding.

1. Place 40 mL of each of the prepared thiosulfate solutions in six different beakers.

2. Starting with the beaker containing the highest concentration of thiosulfate ions, add 5 drops of concentrated hydrochloric acid to each beaker and stir the solution. Note the time that turbidity begins to occur as well as the amount of turbidity.

Safety Information

1. Sodium thiosulfate pentahydrate may be irritating through skin contact and is a strong reducing agent.

2. External contact with HCl causes severe burns, and contact with the eyes may result in a loss of vision. Inhalation causes coughing and choking with possible inflammation and ulceration of the respiratory tract. Work in a fume hood, and wear goggles, gloves, and a rubber apron.

Materials

▲ Sodium thiosulfate pentahydrate, $Na_2S_2O_3 \cdot 5H_2O$: Dissolve the specified amount of solid in 200 mL of distilled water: 12%, 43.5 g; 10%, 36.3 g; 8%, 29.0 g; 6%, 21.8 g; 4%, 14.5 g; and 2%, 7.3 g.

▲ Hydrochloric acid, concentrated HCl: commercial solution.

Concepts

▲ As the concentration of thiosulfate ions, $S_2O_3^{2-}$, decreases, using a constant amount of hydrochloric acid, the amount of time it takes to form colloidal sulfur increases. The formation of colloidal sulfur is observed when the colorless thiosulfate solution starts to become turbid.

▲ Acidification of a sodium thiosulfate solution produces unstable thiosulfuric acid, $H_2S_2O_3$. Decomposition of this acid produces sulfurous acid, H_2SO_3, and colloidal sulfur, noted as cloudiness that develops throughout the solution.

▲ The colloidal suspension of sulfur formed is known as a sol, which is a dispersion of a solid in a liquid phase. Milk of magnesia is another example of a sol.

Reactions

1. $S_2O_3^{2-}(aq) + 2H^+(aq) \rightarrow H_2S_2O_3(aq)$

2. $8H_2S_2O_3(aq) \rightarrow 8H_2SO_3(aq) + S_8(s)$

Notes

A larger, or even smaller, volume of acid could be used, but we prefer 5 drops. This amount allows a comparison between the beakers in terms of time and amount of turbidity.

References

Bacon, E. K. *J. Chem. Educ.* **1948**, *25*, 251–252.

Goldsmith, R. H. *J. Chem. Educ.* **1988**, *65*, 623.

Lamb, W. G. *Sci. Child.* **1984**, 101.

Lippy, J. D., Jr.; Palder, E. L. *Modern Chemical Magic;* Stackpole: Harrisburg, PA, 1959; p 7.

Nebergall, W. H.; Schmidt, F. C.; Holtzclaw, H. F., Jr. *College Chemistry*, 5th ed.; Heath: Lexington, MA, 1976; p 585.

Catalysts in Action

When cobalt(II), iron(III), or copper(II) is added to a solution of tartrate ions and hydrogen peroxide, a dramatic, repeatable color change takes place.

Procedure

Please consult the Safety Information before proceeding.

1. Add 50 mL of 1.0 M sodium potassium tartrate to a tall-form 200-mL beaker.

2. Add 25 mL of 3% hydrogen peroxide and gently warm the solution to 35 °C on a hot plate–stirrer, set at a very slow speed.

3. Add 1–2 mL of 1 M $CoCl_2$ to the solution. Note the color change and the evolution of gas.

4. To repeat, add 25 mL of H_2O_2, and again note the color change.

Safety Information

1. Hydrogen peroxide, 3%, is used as an antiseptic; however, it may cause irritation to mucous membranes, and it may cause serious damage to the eyes.

2. Cobalt(II) chloride hexahydrate, iron(III) chloride hexahydrate, and copper(II) sulfate pentahydrate are all toxic if ingested and may be harmful if they come in contact with mucous membranes or if the dust is inhaled.

3. Sodium potassium tartrate (Rochelle's salt) which has a widely accepted medicinal use, is toxic if ingested and may cause damage if it comes in contact with the mucous membrane of the eyes.

Materials

▲   Copper(II) sulfate penta-hydrate, 1.0 M $CuSO_4 \cdot 5H_2O$: Dissolve 125 g in enough water to make 0.5 L of solution.

▲   Cobalt(II) chloride hexa-hydrate, 0.1M $CoCl_2 \cdot 6H_2O$: Dissolve 112 g in enough water to make 0.5 L of solution.

▲   Iron(III) chloride hexa-hydrate, 1.0 M $FeCl_3 \cdot 6H_2O$: Dissolve 135 g in enough water to make 0.5 L of solution.

▲   Sodium potassium tar-trate tetrahydrate, 1.0 M $NaKC_4H_4O_6 \cdot 4H_2O$: Dissolve 132 g in enough water to make 0.5 L of solution.

▲   Hydrogen peroxide, 3% H_2O_2: Add 5 mL of 30% hy-drogen peroxide to enough water to make 50 mL.

5.   Repeat Steps 1–4 with 1–2 mL of 1 M copper(II) sulfate or iron(III) chloride.

Concepts

▲   The reaction of sodium potassium tartrate and hydrogen peroxide is very slow at best to pro-duce carbon dioxide and possibly oxygen. When cobalt(II) chloride is added, the rapid evolution of a gas (CO_2) is observed.

▲   The pink color of cobalt(II) ion is visible when it is first added to the tartrate–peroxide solu-tion. Immediately, the cobalt is converted to an intermediate, which is green. When the reac-tion is complete—that is, the hydrogen perox-ide is consumed—the pink color of the catalyst (Co^{2+}) returns.

▲   In a slightly different reaction, if copper(II) ions are used, the initial color is blue and then an orange–yellow precipitate is produced. This sus-pended solid is copper(II) oxide. The first reac-tion is one of a catalyst; that is, the copper(II) ion catalyzes the decomposition of the hydro-gen peroxide. Additionally, the tartrate is oxi-dized (CO_2), and copper(II) is reduced to copper(I) and forms Cu_2O. When additional hydrogen peroxide is added, the copper(I) ox-ide is oxidized back to copper(II) ion. This re-action continues until the volume becomes too great; that is, the solution becomes too dilute. Even with this oxidation, the amount of tartrate remains in excess.

▲   When iron(III) is used, the initial solution is yellow and changes to a jet black suspension, probably of iron(II) oxide. This reaction appears to be analogous to the copper reaction. At first, the iron(III) appears to act as a catalyst and then it too becomes a reactant, reduced to the iron(II) state as the oxide. When additional hydrogen peroxide is added, the iron(II) is oxidized back to the iron(III) state where it once again can catalyze the reaction.

▲   Apparently, the tartrate ion reduces both the iron(III) to iron(II) and the copper(II) to copper(I). This reaction may account for the oxidation of the tartrate ion to carbon dioxide and water.

Reactions

1. $C_4H_4O_6^{2-}(aq) + 5H_2O_2(aq) \xrightarrow{Co^{2+},\ Fe^{2+},\ Cu^{2+}}$
$$4CO_2(g) + OH^-(aq) + OH^-(aq) + 6H_2O(l)$$

2. $Co^{2+}(aq) \leftrightharpoons CoC_4H_4O_6(aq\ intermediate)$
  pink   green

3. $2Cu^{2+}(aq) + C_4H_4O_6^{2-}(aq) + 6H_2O_2(aq) \rightarrow$
  blue
$$Cu_2O(s) + 2OH^-(aq) + 4CO_2(g) + 7H_2O(l)$$
           yellow-orange

4. $Fe^{3+}(aq) + C_4H_4O_6^{2-}(aq) + 6H_2O_2(aq) \rightarrow$
  yellow
$$FeO(s) + 2OH^-(aq) + 4CO_2 + 7H_2O(l)$$
         black

5. $Cu_2O(s) + H_2O_2(aq) \rightarrow 2Cu^{2+}(aq) + O_2(g) + H_2O(l)$
  yellow-orange   blue

6. $FeO(s) + H_2O_2(aq) \rightarrow Fe^{3+}(aq) + O_2(g) + H_2O(l)$
  black   yellow

Notes

The copper sulfate reaction is very rapid and may sometimes effervesce over the top of the beaker if the temperature is much above 40 °C. The reactions with iron and cobalt are not as vigorous even at temperatures around 50–60 °C.

References

Deroo, J. *Sci. Teach.* **1974**, *41*, 44.

Ealy, Julie B.; Ealy, James L., Jr. *Close-up on Chemistry;* ACS video manuscript; American Chemical Society: Washington, DC, 1991; pp 20–22.

Lang, M., University of Wisconsin, Stevens Point, WI, unpublished results.

Ruda, P. T. *J. Chem. Educ.* **1978**, *55*, 652.

Sherman, M. C.; Well, D. *J. Chem. Educ.* **1991**, *68,* 1037.

Toth, Z. *J. Chem. Educ.* **1980**, *57*, 464.

Rapid pH Changes with Organic Hydrolysis

When several drops of 2-bromo-2-methylpropane are added to alkaline isopropyl alcohol containing universal indicator, a spectrum of colors is produced.

Procedure

Please consult the Safety Information before proceeding.

1. Place 50 mL of isopropyl alcohol in a tall-form 150-mL beaker on a magnetic stirrer at slow speed.

2. Add 5 drops of NaOH solution.

3. Add 20 drops (or enough to get a deep rich blue color) of universal indicator.

4. Add 5 drops of 2-bromo-2-methylpropane

5. After the indicator has changed to red, add as many drops of NaOH as necessary to return

Safety Information

1. Sodium hydroxide is extremely corrosive to body tissue. Handle with care and avoid all contact with the skin. Inhalation of the dust results in corrosive action on the mucous membranes. Wear gloves when handling.

2. 2-Bromo-2-methylpropane and 2-chloro-2-methylpropane are corrosive and should be handled with gloves. Avoid contact with the skin.

3. Isopropyl alcohol, 70%, is rubbed on the skin for medicinal purposes; however, if ingested it can produce death.

4. Universal indicator is toxic by inhalation.

Materials

▲ Isopropyl alcohol, 50% $(CH_3)_2CHOH$: Add 250 mL of water to 250 mL of 100% isopropyl alcohol. Alternatively, you can use 70% isopropyl alcohol purchased from the drug store and dilute accordingly—357 mL of 70% isopropyl alcohol and 143 mL of water to give a 50% solution. Any brand of fuel-line freeze-up preventative that is 100% isopropyl alcohol could also be used. Be certain it is isopropyl alcohol and not methyl alcohol.

▲ Sodium hydroxide, 0.5 M NaOH: Dissolve 2 g of NaOH to 100 mL of water and store the solution in a dropper bottle.

▲ 2-Bromo-2-methylpropane (98%), C_4H_9Br.

▲ 2-Chloro-2-methylpropane (98%), C_4H_9Cl.

▲ Universal indicator.

the indicator to blue, and a spectrum of colors will splash by again.

6. Repeat Steps 1–5 with 2-chloro-2-methylpropane in place of 2-bromo-2-methylpropane.

Concepts

▲ The reaction is a simple hydrolysis of 2-bromo-2-methylpropane to produce HBr, which causes the solution to be acidic. However, the reaction is slow enough to proceed through a wide range of colors from deep blue (basic) to bright red (acid).

▲ When the reaction has proceeded to the acidic stage, NaOH (base) is added to neutralize the HBr and restore the basic condition of the solution.

▲ Because the hydrolysis will continue until most of the 2-bromo-2-methylpropane is reacted, NaOH can be added many times.

▲ The kinetics of the reaction is first-order for 2-bromo-2-methylpropane. As its concentration decreases, the time to produce acidic (HBr) conditions increases.

▲ Structures are as follows:

2-bromo-2-methylpropane   2-chloro-2-methylpropane

Reactions

1. $CH_3CBr(CH_3)CH_3(aq) + OH^-(aq) \rightarrow$
$CH_3(CH_3)COHCH_3(aq) + Br^-$ [aq (mostly)]

2. $CH_3CCl(CH_3)CH_3(aq) + OH^-(aq) \rightarrow$
$CH_3(CH_3)COHCH_3(aq) + Cl^-$ [aq (mostly)]

Notes

1. An interesting and more colorful variation, but not as dramatic, is to use 5 drops of Congo red indicator along with the universal indicator.

2. The first several times the reaction proceeds, it is fast—10 s. Around the seventh repetition, the times slow to 25–35 s. However, the times are only relative, as the time between the solution's turning red (acidic) and the addition of the NaOH will affect the time it takes to become acidic. The amount of NaOH added will also affect the time for the production of the HBr or HCl to neutralize the base and become acidic.

References

Riley, J. T. *J. Chem. Educ.* **1977**, *54*, 29.

Roberts, J. D.; Caserio, M. C. *Basic Principles of Organic Chemistry;* Benjamin: Reading, MA, 1964; p 312.

Copper Catalyst

When a coil of copper wire is added to a deep purple solution of iron(III) thiosulfate or a deep red solution of iron(III) thiocyanate, the color disappears rapidly.

Procedure

Please consult the Safety Information before proceeding.

1. Place 5 mL of sodium thiosulfate solution in each of four test tubes.

2. Place 2 drops of potassium thiocyanate solution in the third and fourth tubes. Stopper and shake the tubes to mix the solution thoroughly.

3. Place 5 mL of iron(III) chloride in each tube. Shake the tubes to mix the solution thoroughly. Notice the purple color formed in the first two

Safety Information

1. Nitric acid is corrosive to all body tissues. Inhalation of the concentrated vapor may cause serious lung damage; contact with the eyes may result in a total loss of vision. Continued exposure to the vapor of nitric acid may cause chronic bronchitis. Work in a fume hood, and wear gloves, a rubber apron, and goggles.

2. Sodium thiosulfate pentahydrate is moderately toxic by ingestion and mildly irritating to the skin and tissues.

→

3. Iron(III) chloride hexa-hydrate is toxic by ingestion and a strong irritant to the skin and tissue.

4. Potassium thiocyanate is very toxic by ingestion. When heated, it emits poisonous cyanide fumes. Although you are not heating this substance, you must practice prudent laboratory safety to prevent contamination of the laboratory space. Wear gloves, a laboratory coat, and eye protection.

tubes and the blood-red color in the last two tubes.

4. Immediately place a coil of copper wire into the solutions in the first and third tubes. Shake the tubes. Notice the rapid color change. The solutions in these test tubes become pale yellow.

5. The second and fourth test tubes are used as references to show the noncatalyzed slow change of color (2–5 min).

Concepts

▲ When iron(III) ions come in contact with thio-sulfate ions, $S_2O_3^{2-}$, a slow redox reaction results. Iron(III) is reduced to iron(II) and thiosulfate is oxidized to tetrathionate, $S_4O_6^{2-}$.

▲ An intermediate purple complex, $Fe(S_2O_3)_2^-$ is formed between iron(III) and thiosulfate ions. In the presence of iron(III) ions, the purple color of the iron(III) thiosulfate fades slowly (2–5 min) with the formation of iron(II) and dithionate ions.

▲ Also, the red complex ion, iron(III) thiocyanate, $FeSCN^{2+}$, forms. Iron(III) thiocyanate does not dissociate to any appreciable extent. The few iron(III) ions that do form are reduced to iron(II) as quickly as they dissociate. In the presence of iron(III) ions and thiosulfate, the red color of the complex fades very slowly (2–5 min).

▲ When copper(II) ions are added to a solution of iron(III) and thiosulfate ions, the redox reaction is speeded up considerably. The reaction involves copper(II) reacting with the iron(III) complex to produce copper(I), iron(II), and dithionate ions. Copper(I) also reacts with iron(III) to produce copper(II) and iron(II). With copper(II) ions acting as a catalyst, the purple color in the first and the red color in the third test tube change almost immediately.

▲ When copper metal is placed in the solution, the very few copper ions present on the surface of the metal are enough to catalyze the reaction.

Reactions

1. $2Fe^{3+}(aq) + 2S_2O_3^{2-}(aq) \rightarrow 2Fe^{2+}(aq) + S_4O_6^{2-}(aq)$

2. $Fe^{3+}(aq) + 2S_2O_3^{2-}(aq) \rightarrow Fe(S_2O_3)_2^{-}(aq)$
   $\qquad\qquad\qquad\qquad\qquad\qquad$ purple

3. $Fe(S_2O_3)_2^{-}(aq) + Fe^{3+}(aq) \rightarrow$
   $\qquad\qquad\qquad 2Fe^{2+}(aq) + S_4O_6^{2-}(aq)$ slow

4. $Fe^{3+}(aq) + SCN^{-}(aq) \rightarrow FeSCN^{2+}(aq)$
   $\qquad\qquad\qquad\qquad\qquad$ red

5. $Fe^{3+}(aq) + 2S_2O_3^{2-}(aq) + FeSCN^{2+}(aq) \rightarrow$
   $\qquad 2Fe^{2+}(aq) + SCN^{-}(aq) + S_4O_6^{2-}(aq)$ slow

6. Copper(II) catalyst:

   $Fe(S_2O_3)_2^{-}(aq) + Cu^{2+}(aq) \rightarrow$
   $\qquad\qquad Cu^{+}(aq) + Fe^{2+}(aq) + S_4O_6^{2-}(aq)$ rapid

   $FeSCN^{2+}(aq) + Cu^{2+}(aq) + 2S_2O_3^{2-}(aq) \rightarrow$
   $Cu^{+}(aq) + Fe^{2+}(aq) + S_4O_6^{2-}(aq) + SCN^{-}(aq)$ rapid

   $Cu^{+}(aq) + Fe^{3+}(aq) \rightarrow Cu^{2+}(aq) + Fe^{2+}(aq)$ rapid

Materials

▲ Iron(III) chloride hexa-hydrate, 0.1 M $FeCl_3 \cdot 6H_2O$: Dissolve 2.7 g in enough water to make 100 mL of solution.

▲ Sodium thiosulfate penta-hydrate, 0.1 M $Na_2S_2O_3 \cdot 5H_2O$: Dissolve 2.4 g in enough water to make 100 mL of solution.

▲ Copper(II) sulfate penta-hydrate, 0.1 M $CuSO_4 \cdot 5H_2O$: Dissolve 2.5 g in enough water to make 100 mL of solution.

▲ Potassium thiocyanate, 0.1 M KSCN: Dissolve 1.0 g in enough water to make 100 mL of solution.

▲ Copper wire.

Notes

1. The copper wire should be cleaned in 6 M nitric acid and rinsed thoroughly with distilled water. This action will not prove that you are not adding copper ions, but it is a suitable attempt to show that the copper metal can catalyze the reaction.

2. In the recent past, this reaction was a qualitative test for the presence of very small amounts of copper ions. Because it causes the fading of the red color in an iron thiocyanate solution, as little as 0.2 parts per million of copper can be detected.

References

Feigl, F. *Spot Tests, Vol. I, Inorganic Chemistry*, 4th English translation; Elsevier: Houston, TX, 1954; p 205.

Greenwood, N. N.; Earnshaw, A. *Chemistry of the Elements;* Pergamon: New York, 1984; pp 1263–1271.

Acids and Bases

Determination of Dissociation Constants Using Current

The dissociation constants, K_a of a weak acid or K_b of a weak base, can be calculated by measuring the current, in milliamperes, in the weak acid or base and comparing it with the current in a range of concentrations of a strong acid (see Investigation 77).

Procedure

Please consult the Safety Information before proceeding.

1. Using a multimeter, place a red patch cord in "A" and a black one in "COM". Attach the red patch cord to the red wire of the light-emitting diode (LED) apparatus (consult Investigation 77) using an alligator clip, and attach the black patch cord to the black wire of the LED apparatus.

2. Set the multimeter on 200 mA. If necessary, press the DC/AC button so the screen shows "AC".

3. Chloroacetic acid may be fatal if swallowed or absorbed through the skin. The material is destructive to the tissue of the mucous membranes and the upper respiratory tract, eyes, and skin.

4. Inhalation of concentrated ammonia causes edema of the respiratory tract, spasm of the glottis, and asphyxia. Treatment must be prompt to prevent death.

5. Diethylamine is a severe eye, skin, and mucous membrane irritant and is also toxic.

3. Place 40 mL of acid or base in a 50-mL beaker.

4. Plug in the adapter.

5. Insert the probes of the LED apparatus in the solution for 10 s to a specific height (see Notes). Record the current.

6. Rinse the probes with distilled water and wipe the probe dry. Determine the current two more times, washing and drying the electrodes between trials. Average the results of the three trials.

7. Repeat Steps 3–6 using the other solutions.

8. Using the graph of concentration versus current from Investigation 77, determine the concentration of HCl that corresponds to the current value for the solution. Calculate K_a or K_b.

Concepts

▲ Hydrochloric acid, HCl, ionizes essentially 100% in water and reacts to form hydronium ions, H_3O^+, and chloride ions, Cl^-. If the concentration of the acid is known, then the concentration of hydronium ions is known and is the same as that of the strong acid.

▲ In contrast to hydrochloric acid in dilute solution, a weak acid (or base) is ionized only a few percent at the same concentration. The concentration of the ions in a 1:1 weak acid (or base) solution is equal to that in a hydrochloric acid solution that has been diluted enough to pass an equal electric current under the same experimental conditions. This fact is used to measure the concentrations of ions in weak acid and base solutions.

▲ Weak acids or bases are not 100% ionized but undergo only partial ionization. Equations for the ionization of weak acids or bases are written with forward and reverse arrows. Such an equation acknowledges that molecular acid, hydronium ions, and the anion (negative ion), are present at equilibrium for a weak acid. For weak bases, molecular base, the cation (positive ion), and the anion (negative ion) are present.

▲ If 0.00650 M HCl has the same current as 2.00 M CH_3COOH, then the concentration of H_3O^+ and

CH_3COO^- is 0.00650 M, because the concentration of H_3O^+ ions for HCl is 0.00650 M. The concentration of nonionized acetic acid is 2.00 M – 0.00650 M = 1.994 M.

$$K_a = \frac{[H_3O^+]\,[CH_3COO^-]}{[CH_3COOH]} = \frac{(0.00650\ \text{M})\ (0.00650\ \text{M})}{1.994\ \text{M}} = 2.12 \times 10^{-5}$$

▲ The ionization constant for a weak acid is represented by K_a and for a weak base, K_b. Calculation of K_a for a weak acid, such as acetic acid, is based on the concentration of hydronium ions times the concentration of acetate ions divided by the concentration of the weak nonionized acetic acid. K_a or K_b, the equilibrium or ionization constant, must be experimentally determined.

▲ As found in *Lange's Handbook of Chemistry*, equilibrium values for the acids and bases at 25 °C follow, as well as experimental values based on the results:

Compound	Equilibrium Constant	Experimental
Acetic acid	1.75×10^{-5}	2.28×10^{-5}
Oxalic acid (K_{a1})	5.36×10^{-2}	2.39×10^{-2}
Chloroacetic acid	1.36×10^{-3}	1.34×10^{-3}
Ammonia	1.81×10^{-5}	1.81×10^{-5}
Diethylamine	5.81×10^{-4}	1.66×10^{-4}

▲ Following are our results based on the device used:

Substance	Current (mA)	HCl (M)	Experimental
2.00 M acetic acid	20.3	0.00650	$K_a = 2.12 \times 10^{-5}$
1.00 M acetic acid	17.6	0.00550	$K_a = 3.04 \times 10^{-5}$
0.100 M acetic acid	7.7	0.00150	$K_a = 2.28 \times 10^{-5}$
1.00 M oxalic acid	34.0	0.143	$K_a = 2.39 \times 10^{-2}$
1.00 M chloroacetic acid	31.2	0.0360	$K_a = 1.34 \times 10^{-3}$
1.00 M ammonia	15.2	0.00425	$K_b = 1.81 \times 10^{-5}$
1.00 M diethylamine	26.5	0.0128	$K_b = 1.66 \times 10^{-4}$

Materials

▲ For all solutions, add the specified amount of stock solution of acid or base to enough distilled water to make 100 mL of solution:

a. Acetic acid, 2.00 M CH_3COOH: 11.50 mL.
b. Acetic acid, 1.00 M CH_3COOH: 5.75 mL.
c. Acetic acid, 0.100 M CH_3COOH: 0.575 mL.
d. Oxalic acid dihydrate, 1.00 M $HOOCCOOH \cdot 2H_2O$: 12.61 g.
e. Chloroacetic acid, 1.00 M $ClCH_3COOH$: 9.45 g.
f. Ammonia, 1.00 M NH_3: 6.76 mL.
g. Diethylamine, 1.00 M $(C_2H_5)_2NH$: 10.30 mL.

▲ Multimeter.
▲ LED apparatus.

Reactions

1. $HCl(aq) + H_2O(l) \rightarrow H_3O^+(aq) + Cl^-(aq)$

2. $CH_3COOH(aq) + H_2O(l) \leftrightarrows H_3O^+(aq) + CH_3COO^-(aq)$

3. $HOOCCOOH(aq) + H_2O(l) \leftrightarrows$
$$H_3O^+(aq) + HOOCCO^-(aq)$$

4. $ClCH_2COOH(aq) + H_2O(l) \leftrightarrows$
$$H_3O^+(aq) + ClCH_2COO^-(aq)$$

5. $NH_3(aq) + H_2O(l) \leftrightarrows NH_4^+(aq) + OH^-(aq)$

6. $(C_2H_5)_2NH(aq) + H_2O(l) \leftrightarrows$
$$(C_2H_5)_2NH_2^+(aq) + OH^-(aq)$$

Notes

1. The choice of 40 mL of solution and a 50-mL beaker is somewhat arbitrary. What is necessary, though, is that the same volume of solution in the same size beaker be used each time and the probes submerged the same distance (consult Notes in Investigation 77).

2. The solutions can be used over again as long as the probes are always rinsed and dried to avoid contamination of the solutions. Store the solutions in labeled bottles.

3. The experimental results are not meant to agree quantitatively with results found in *Lange's Handbook of Chemistry*, but the demonstration is meant to provide a means by which similar results can be obtained, and K_a or K_b can then be calculated by your students.

References

Dean, J. A. *Lange's Handbook of Chemistry*, 13th ed.; McGraw-Hill: New York, 1985; pp 5-18–5-52.

Sillen, L. G. *Stability Constants of Metal-Ion Complexes;* Chemical Society: London, 1964; pp 149, 360, 364, 374.

A Colorful Glass Tube

A glass tube, stoppered at both ends, is filled with a purple cabbage juice solution. Adding hydrochloric acid to one end and sodium hydroxide solution to the other end and then inverting the tube produces a gradation of red, purple, blue, and green colors.

Procedure

Please consult the Safety Information before proceeding.

1. Use a glass tube stoppered at both ends. Fill the tube with cabbage juice solution, leaving a small space at one end.

2. Using a pipet, deliver 2 mL of hydrochloric acid solution in one end. Stopper the tube and invert.

3. Remove the other stopper. Using another pipet, deliver 2 mL of sodium hydroxide solution in this end. Stopper the tube.

Safety Information

1. External contact with hydrochloric acid causes severe burns, and contact with the eyes may result in a loss of vision. Inhalation causes coughing and choking with possible inflammation and ulceration of the respiratory tract. Work in a fume hood, and wear goggles, gloves, and a rubber apron.

2. Sodium hydroxide is corrosive to all tissues. Inhalation of the dust or concentrated mist may cause damage to the respiratory tract. Wear gloves, an apron, and goggles when handling.

Materials

▲ Cabbage juice: Cut up a red cabbage into small pieces and boil in distilled water for 1 h. Decant the purple solution and use it for the demonstration.

▲ Hydrochloric acid, 0.1 M HCl: Adding acid to water, mix 8.3 mL of concentrated acid with enough water to make 1000 mL of solution.

▲ Sodium hydroxide, 0.1 M NaOH: Dissolve 0.40 g in enough water to make 100 mL of solution.

4. Invert the tube and observe the colors.

5. Invert the tube again and observe the colors.

Concepts

▲ Plant pigments occur in plastids or in solution in the sap of plants. One of the classes of "sap pigments" is the anthocyanins, which are present in red cabbage. The purple color that results from boiling red cabbage in water is a mixture of anthocyanins.

▲ Anthocyanins are red in acid, violet when neutral, and vary from dull red or reddish brown to purple and green in basic solution.

▲ Anthocyan pigments are present in flowers, fruits, and leaves of plants in the form of glucosides. By hydrolysis, anthocyanins are converted into glucose or other monosaccharides and colored anthocyanidins. The three parent anthocyanidins are named pelargonidin, cyanidin, and delphinidin:

pelargonidin chloride   cyanidin chloride   delphinidin chloride

▲ The number of hydroxyl groups in the anthocyanidin part of the molecule, the other pigments present, and the acidity of the cell sap determine the various shades of natural colors.

▲ If the glass tube is clamped in place, the process of diffusion will occur as the colors in the glass tube mix with each other, until the final color is purple.

▲ At pH 2, cyanin, one of the pigments in red cabbage juice solution, has broad absorption bands in the ultraviolet region of the electromagnetic spectrum and in the 400–600-nm wavelength range of visible light. Blues and greens are absorbed, and reds and oranges are transmitted.[1]

▲ With an increase in pH to 7, most of the absorbance is from 540 to 640 nm, where orange, yellow, and green are absorbed. The transmission of reds and blues results in the observed purple color.[1]

▲ When a pH of 12 is reached, there is absorption in the near infrared and of longer wavelengths in the visible range. Energy is transmitted in the blue part of the spectrum as well as a narrow band in the 580–600-nm yellow range. This combination of blue and yellow results in green.[1]

Reactions

The reaction, using cyanin with its representative structures in acid and base, is

purple, pH 7

red, pH 2

green, pH 11

[1]Rand, R. Personal communication, August 25, 1993, Albuquerque Academy, Albuquerque, NM.

Notes

1. The glass tube can be different sizes. We used one about 40 cm long with an inner diameter of 13 mm, as well as one about 95 cm long with an inner diameter of 35 mm. By far, the larger one was better, although both worked fine. The size of your group might dictate which one to use.

2. If you would like to do the demonstration with only household products, use vinegar for the acid and a window cleaner for the base.

References

Becker, R. "Acid/Base Magic Wand"; Chemical Education '91 handout; University of Wisconsin, Oshkosh, WI, 1991.

Fuson, R. C.; Connor, R.; Price, C. C.; Snyder, H. R. *A Brief Course in Organic Chemistry;* Wiley: London, 1941; pp 167–168.

Chemistry and Biochemistry of Plant Pigments; Goodwin, J. B., Ed.; Academic: New York, 1965.

Kemp, D. S.; Vellaccio, F. *Organic Chemistry;* Worth: New York, 1980; pp 983–984.

Kingzett, C. T. *Chemical Encyclopedia;* Van Nostrand: New York, 1928; pp 565–567.

Lowy, A.; Harrow, B.; Apfelbaum, P. M. *Introduction to Organic Chemistry;* Wiley: New York, 1945; p 385.

Nassau, K. *The Physics and Chemistry of Color;* Wiley: New York, 1983; p 132.

Wertheim, E. *Textbook of Organic Chemistry;* Blakiston: Philadelphia, PA, 1939; pp 650–652.

Neutralization of Cabbage Juice

Addition of hydrochloric acid to a purple cabbage juice solution makes it red. Addition of sodium hydroxide solution to another sample of cabbage juice solution makes it green. Addition first of hydrochloric acid and then of sodium hydroxide solution to the purple solution keeps it purple.

Procedure

Please consult the Safety Information before proceeding.

1. Place 10 mL of cabbage solution in each of four different test tubes. One test tube is a reference.

2. Place 5 drops of hydrochloric acid in the first test tube. Swirl the contents to mix. A red color results.

3. Place 5 drops of sodium hydroxide in the second test tube. Swirl the contents to mix. A green color results.

Materials

▲ Cabbage juice: Cut up red cabbage into small pieces and boil in distilled water for 1 h. Decant the purple solution and use it for the demonstration.

▲ Hydrochloric acid, 1.0 M HCl: Adding acid to water, mix 8.3 mL of concentrated acid with enough water to make 100 mL of solution.

▲ Sodium hydroxide, 1.0 M NaOH: Dissolve 4.0 g in enough water to make 100 mL of solution.

4. Place 5 drops of hydrochloric acid in the third test tube. Swirl the contents to mix. Now add 5 drops of sodium hydroxide solution. Swirl the contents to mix. The purple color of the cabbage juice alone is produced. Compare with the reference tube.

Concepts

▲ See Investigation 59 for a discussion of plant pigments.

▲ When a strong acid such as HCl is mixed with a strong base such as NaOH, a neutralization reaction occurs. Because the color of red cabbage in water is purple, adding equal volumes of equal concentrations of acid and base restores the original purple color.

Reactions

1. Neutralization reaction:

$$H_3O^+(aq) + OH^-(aq) \leftrightarrows 2H_2O(l)$$

2. The reaction, using cyanin with its representative structures, is shown in Investigation 59.

Notes

1. It is important to prepare the solutions as carefully as possible and count the drops accurately. Any equal volumes of acid and base could be used, but you must be exact. If you wish to use an equal number of milliliters of acid and base—for example, 2 mL of each—use a pipet or an automatic pipetter.

2. Check the pH of the cabbage juice solution before beginning the demonstration to make sure it is neutral. If not, adjust the pH with any needed acid or base.

3. The demonstration works well by cooking the cabbage in water and just using the aqueous solution. It is not necessary to add alcohol.

References

Fuson, R. C.; Connor, R.; Price, C. C.; Snyder, H. R. *A Brief Course in Organic Chemistry;* Wiley: London, 1941; pp 167–168.

Kemp, D. S.; Vellaccio, F. *Organic Chemistry;* Worth: New York, 1980; pp 983–984.

Kingzett, C. T. *Chemical Encyclopedia;* Van Nostrand: New York, 1928; pp 565–567.

Lowy, A.; Harrow, B.; Apfelbaum, P. M. *Introduction to Organic Chemistry;* Wiley: New York, 1945; p 385.

Wertheim, E. *Textbook of Organic Chemistry;* Blakiston: Philadelphia, PA, 1939; pp 650–652.

Appearing and Disappearing *o*-Cresolphthalein

A colorless solution of hydrochloric acid and indicator, *o*-cresolphthalein, is made purple with a few drops of sodium hydroxide. Continued addition of sodium hydroxide eventually makes the purple solution colorless.

Procedure

Please consult the Safety Information before proceeding.

1. Place 10 mL of hydrochloric acid solution in a 500-mL beaker on a magnetic stirrer. Turn on the stirrer.

2. Add 1.5 mL of *o*-cresolphthalein indicator. The mixture is colorless.

3. Add sodium hydroxide solution dropwise until a purple color is obtained—about 8 drops.

4. Thereafter, continue to slowly add sodium hy-

Safety Information

1. External contact with hydrochloric acid causes severe burns, and contact with the eyes may result in a loss of vision. Inhalation causes coughing and choking with possible inflammation and ulceration of the respiratory tract. Work in a fume hood and wear goggles, gloves, and a rubber apron.

2. Sodium hydroxide is corrosive to all tissues. Inhalation of the dust or concentrated mist may cause damage to respiratory tract. Wear gloves, an apron, and goggles when handling.

Materials

▲ Hydrochloric acid, 0.5 M HCl: Slowly adding concentrated acid to water, mix 4.2 mL with enough distilled water to make 100 mL of solution.

▲ Sodium hydroxide, 6.0 M NaOH: Dissolve 120 g in enough distilled water to make 500 mL of solution.

▲ o-Cresolphthalein indicator: Dissolve 0.04 g in enough ethanol to make 100 mL of solution.

droxide solution until the solution is again colorless—about 450 mL. The magnetic stirrer will help mix the solutions quickly.

Concepts

▲ When a strong acid and a strong base are mixed, a neutralization reaction takes place, with the overall net reaction producing water. In this reaction, neutralization occurs when the indicator turns purple.

▲ The pH range for the indicator, o-cresolphthalein, is 8.2–9.8, going from colorless to purple. The structure for the indicator in its colorless form is represented by compound 1. When base is added, the lactone ring of the indicator is broken and the lactoid structure shown by compound 2 results. The characteristic color, purple, of the triphenylmethane dye is produced.

▲ Continued addition of concentrated NaOH results in the conversion of the triphenylmethane dye to the colorless carbinol form shown by compound 3. This reaction occurs above pH 9.8.

Reactions

Neutralization: $H_3O^+(aq) + OH^-(aq) \leftrightarrows 2H_2O(l)$

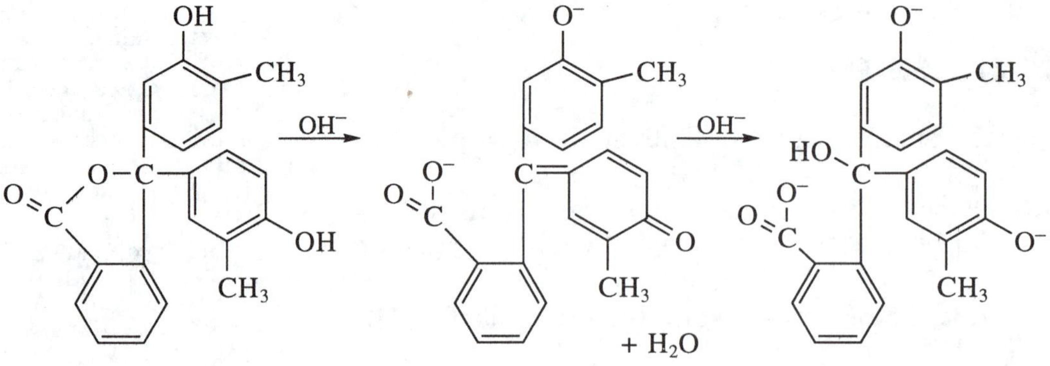

1   colorless   2   purple   3   colorless

Notes

Phenolphthalein, with a pH range of 8.2–10.0, can also be used. Only about 55 mL of base will be needed. Although thymolphthalein has a similar pH range, 9.4–10.6, even at 1150 mL of base, a faint blue color is still seen at the concentrations of hydrochloric acid and sodium hydroxide used in this demonstration.

References

Lippy, J. D., Jr.; Palder, E. L. *Modern Chemical Magic*; Stackpole: Harrisburg, PA, 1959; p 9.

Wittke, G. *J. Chem. Educ.* **1983**, *60*, 239.

Blue to Pink and Pink to Blue

A colorless solution is poured into a beaker containing two solids, and a blue solution called Prussian blue is formed. When the contents of this beaker are poured into a third beaker, a pink solution results because of a reaction between phenolphthalein and a basic solution of sodium hydroxide. In a second demonstration, a pink solution results from a reaction between phenolphthalein and a basic solution of sodium hydroxide. Pouring the contents of this beaker into one containing two solids results in the formation of a blue solution.

Procedure 1: Blue to Pink

Please consult the Safety Information before proceeding.

1.	Label three 250-mL beakers with the numbers 1, 2, and 3. In beaker 1, place 100 mL of H_2O

Safety Information

1.	There is insufficient toxicity data on phenolphthalein, but use it with caution.

2.	Potassium hexacyanoferrate(II) trihydrate is an eye and skin irritant. It does not appear to release the cyanide ion.

3.	Iron(II) ammonium sulfate hexahydrate may be harmful by inhalation, ingestion, or skin absorption. It causes eye and skin irritation. It is irritating to mucous membranes and the upper respiratory tract. →

4. External contact with hydrochloric acid causes severe burns, and contact with the eyes may result in loss of vision. Inhalation causes coughing and choking with possible inflammation and ulceration of the respiratory tract.

5. Sodium hydroxide is harmful if swallowed, inhaled, or absorbed through the skin. The material is extremely destructive to tissue of the mucous membranes and the upper respiratory tract, eyes, and skin.

and 2 mL of phenolphthalein. In beaker 2, place a few grains of both iron(II) ammonium sulfate hexahydrate and potassium hexacyanoferrate(II) trihydrate. In flask 3, place 100 mL of H_2O and 20 mL of 6.0 M NaOH.

2. Pour beaker 1 into beaker 2 and mix; blue color is produced.

3. Pour beaker 2 into beaker 3 and mix; pink color is produced.

Procedure 2: Pink to Blue

1. Label three 250-mL beakers with numbers 1, 2, and 3. In beaker 1, place 100 mL of H_2O, 2 mL of 6.0 M HCl, and 10 drops of phenolphthalein. In beaker 2, place 2 mL of 6.0 M NaOH. In beaker 3, place 0.5 g (1/4 tsp) of each of the two solids, iron(II) ammonium sulfate hexahydrate and potassium hexacyanoferrate(II) trihydrate.

2. Pour beaker 1 into beaker 2 and mix; pink color is produced.

3. Pour beaker 2 into beaker 3 and mix; blue color is produced.

Concepts

▲ The blue compound produced in beaker 2 under Blue to Pink is called Prussian blue. Iron(II), Fe^{2+}, combines with cyanide ions, CN^-, forming the complex ion $[Fe(CN)_6]^{4-}$, called hexacyanoferrate(II). Fe^{2+} undergoes oxidation by oxygen in the air, forming Fe^{3+}. Iron(III), Fe^{3+}, combines with hexacyanoferrate(II), forming Prussian blue.

▲ Under Blue to Pink, base is present in beaker 3, and when the contents of beakers 2 and 3 are mixed, a reaction takes place with phenolphthalein indicator, which turns pink in the presence of hydroxide ions, OH^-.

▲ Under Pink to Blue, a pink color results when beakers 1 and 2 are mixed because of the presence of OH^- and phenolphthalein indicator in beaker 2. The acid present in beaker 1 does not affect the indicator, because phenolphthalein is colorless in acid. The presence of the acid helps the Prussian blue, which is produced when beakers 2 and 3 are mixed, to look blue, not purple.

▲ When the concentration of hydroxide ion, OH^-, increases, the equilibrium shifts to the product side of the phenolphthalein indicator reaction, where the indicator color is pink.

Reactions

1. General indicator reaction:

$$HIn(aq) + OH^-(aq) \rightleftharpoons H_2O(l) + In^-(aq)$$

colorless   pink

2. Specific indicator reaction:

colorless   purple

3. Formation of Prussian blue:

$$Fe^{2+}(aq) + 6CN^-(aq) \rightarrow [Fe(CN)_6]^{4-}(aq)$$

hexacyanoferrate(II)

$$Fe^{3+}(aq) + K^+(aq) + [Fe(CN)_6]^{4-}(aq) + H_2O(l) \rightarrow$$

(from oxidation $KFe[Fe(CN)_6] \cdot H_2O(s)$
of Fe^{2+} in the air)   Prussian blue

Notes

1. The reaction can be demonstrated by using iron(III) ammonium sulfate, but we think it works much better with the iron(II).

Materials

▲   Sodium hydroxide, 6.0 M NaOH: Dissolve 24 g in enough distilled water to make 100 mL of solution.

▲   Hydrochloric acid, 6.0 M HCl: Adding concentrated acid to water, mix 50 mL with enough distilled water to make 100 mL of solution.

▲   Phenolphthalein indicator: Dissolve 1 g in 99 mL of ethanol.

▲   Iron(II) ammonium sulfate hexahydrate, $Fe(NH_4)_2(SO_4)_2 \cdot 6H_2O$, powder.

▲   Potassium hexacyanoferrate(II) trihydrate, $K_4Fe(CN)_6 \cdot 3H_2O$, powder.

2.   If for some reason the pink under Blue to Pink is purple, adjust the solution pH by increasing the volume of base in beaker 3. Likewise, if the blue is too purple in beaker 3 under Pink to Blue, adjust the solution pH by increasing the volume of acid in beaker 1.

3.   If there are any chunks of solid, grind them with a mortar and pestle.

References

Greenwood, N. N.; Earnshaw, A. *Chemistry of the Elements;* Pergamon: New York, 1984; p 1271.

Hansen, L. D.; Litchman, W. M.; Daub, G. H. *J. Chem. Educ.* **1969,** *46,* 46.

Kohn, M. *J. Chem. Educ.* **1943,** *20,* 198.

Lippy, J. D., Jr.; Palder, E. L. *Modern Chemical Magic;* Stackpole: Harrisburg, PA, 1959; p 7.

Nebergall, W. H.; Schmidt, F. C.; Holtzclaw, H. F., Jr. *College Chemistry*, 5th ed.; Heath: Lexington, MA, 1976; p 909.

Phenolphthalein Is Orange in Acid and Colorless in Base

Phenolphthalein is shown to have a bright orange color in strong acid and to be colorless in a strong base, contrary to popular belief.

Procedure

Please consult the Safety Information before proceeding.

1. Add 5 mL of 1 M sodium hydroxide to one test tube and 5 mL of 1 M sulfuric acid to a second test tube. Add 5-6 drops of phenolphthalein indicator to each tube. The color of the base should be the familiar pink and that of the acid, colorless.

2. Add approximately 2-3 mL of concentrated sulfuric acid to a small test tube, and carefully add drops of phenolphthalein solution until the solution shows a distinct orange coloration.

Safety Information

1. Sulfuric acid is corrosive to all body tissues. Inhalation of concentrated vapor may cause serious lung damage; contact with the eyes may result in a total loss of vision.

2. Sodium hydroxide is corrosive to all tissues. Inhalation of the dust or concentrated mist may cause damage to the respiratory tract. Wear gloves, an apron, and goggles when handling.

Materials

▲   Phenolphthalein indicator, 1.0%: Add 1.0 g to enough 95% ethanol to make 100 mL of solution.

▲   Sulfuric acid, 98% H_2SO_4: stock commercial solution.

▲   Sodium hydroxide, 1 M NaOH: Dissolve 4 g in enough distilled water to make 100 mL of solution.

▲   Sulfuric acid, 1 M H_2SO_4: Carefully add 6 ml of concentrated acid to 94 ml of distilled water and stir the solution.

3.   To another test tube add 2–3 mL of 1 M sodium hydroxide solution and phenolphthalein until the solution turns deep pink. Then add sodium hydroxide (approximately five or six pellets) and shake the tube. The solution will slowly become colorless in the vicinity of the pellets. Eventually the entire solution will become colorless.

Concepts

▲   In concentrated H_2SO_4, the γ-lactone ring of phenolphthalein is opened after protonation and then forms the triphenylmethyl cation.

▲   In concentrated NaOH, the triphenylmethane dye is converted to the colorless carbinol by the addition of a hydroxyl group.

▲   The lactoid form of phenolphthalein is stable in solutions of pH less than 8. In solutions of pH greater than 8, the γ-lactone ring is opened to form the deep pink chromophore of the triphenylmethane dye.

Reactions

phenolphthalein   1
(colorless)

phenolphthalein   2
(purple)

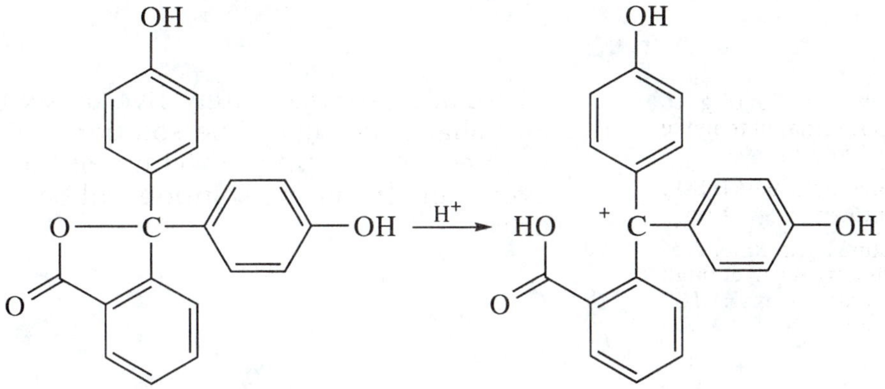

phenolphthalein 1
(colorless)

phenolphthalein 3
(orange)

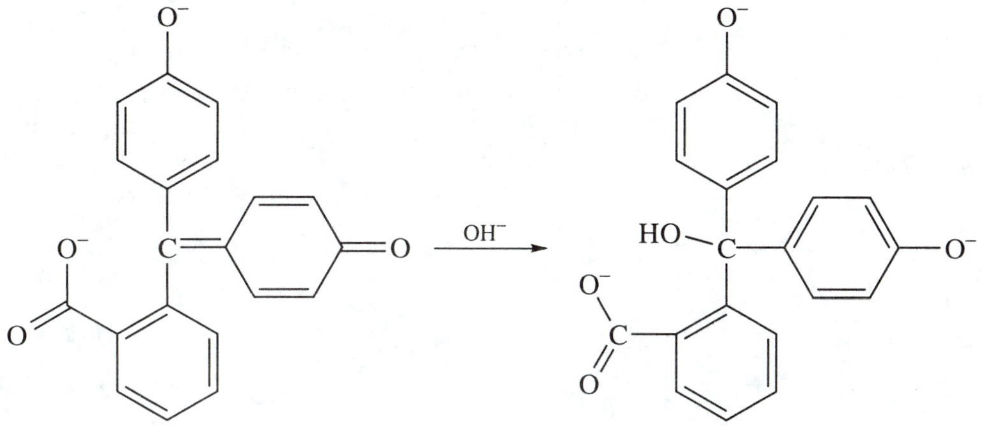

phenolphthalein 2
(purple)

phenolphthalein 4
(colorless)

References

McCullock, L. *J. Chem. Educ.* **1946**, *23*, 473.

Rolf, F. *J. Chem. Educ.* **1964**, *41*, A121.

Wittke, G. *J. Chem. Educ.* **1983**, *60*, 239.

Copper Reacts with Hydrochloric Acid

A hot yellow solution containing tetrachloro-cuprate(II) ions is chilled. Upon addition of color-less ammonia, a blue solution containing tetra-amminecopper(II) ions results.

Procedure

Please consult the Safety Information before pro-ceeding.

1. Place 1/4 tsp (5–10 g) of granular copper in a test tube. Add 5 mL of concentrated hydrochloric acid.

2. Decant the yellow solution. Discard it.

3. Add another 5 mL of concentrated hydrochloric acid. Heat the solution in a boiling water bath for 5 min.

4. Decant the yellow solution into another test tube. Chill the solution in an ice bath for 2 min.

Materials

▲ Hydrochloric acid, concentrated HCl: stock commercial solution.

▲ Copper, granular.

▲ Ammonia, aqueous, concentrated NH_3: stock commercial solution.

5. Carefully add concentrated aqueous ammonia dropwise until a blue color appears.

Concepts

▲ Copper does not react at room temperature with 1.0 M hydrochloric acid. The standard electromotive force is –0.34 V for the overall reaction:

$$Cu^0 + 2H^+ \rightarrow Cu^{2+} + H_2$$

Two half-reactions contribute to the reaction:

$$Cu^0 \rightarrow Cu^{2+} + 2e^- \qquad -0.34 \text{ V}$$
$$2H^+ + 2e^- \rightarrow H_2 \qquad 0.00 \text{ V}$$

A negative reduction potential means that the reaction will not occur spontaneously at room temperature without additional input of energy. With hot concentrated 12 M HCl, the reaction will take place.

▲ As shown in the reaction, copper is oxidized to copper(II) by losing two electrons. It acts as the reducing agent. The two hydrogen ions are reduced to hydrogen by gaining two electrons. They act as the oxidizing agent.

▲ The names and the molecular shapes of the two complex ions are:

Ion	Name	Shape
$CuCl_4^{2-}$	tetrachlorocuprate(II)	square planar
$Cu(NH_3)_4^{2+}$	tetraamminecopper(II)	square planar

Reactions

1. $Cu(s) + 2H^+(aq) + 4Cl^-(aq) \rightarrow CuCl_4^{2-}(aq) + H_2(g)$
   hot,   yellow
   concentrated

2. $CuCl_4^{2-}(aq) + 4NH_3(aq) \rightarrow$
   yellow

                          $Cu(NH_3)_4^{2+}(aq) + 4Cl^-(aq)$
                                blue

Notes

1. Granular copper or copper turnings work best. Copper powder tends to clump and not react well at all.

2. Use only fresh commercial concentrated hydrochloric acid. Concentrated acid is very hygroscopic and will not work well if not fresh.

References

Nebergall, W. H.; Schmidt, F. C.; Holtzlaw H. F., Jr. *College Chemistry,* 5th ed.; Heath: Lexington, MA, 1976; pp 871–872.

Walker, N.; George, D. L. *J. Chem. Educ.* **1968**, *45*, A429.

A Blue Food Coloring That Is Also Green and Yellow

When a strong acid is added to blue food coloring (FD&C Blue No. 1), the color changes from blue to green to yellow.

Procedure

Please consult the Safety Information before proceeding.

1. Place 5 mL of each of the following concentrations of HCl into eight test tubes: 6.0, 5.0, 4.0, 3.0, 2.0, 1.0, 0.5, and 0.1 M.

2. Add 1 drop of FD&C Blue No. 1 food coloring to each test tube and shake the tube. Observe the color change from deep blue to bright yellow.

Concepts

▲ The FD&C Blue No. 1 food coloring is a complex molecule and reacts to hydrogen ions like other more familiar acid–base indicators.

▲ The blue color is stable for a pH of 2–14. Below pH 2, the color begins to change to greenish. Some of the yellow form of the indicator is present, and it mixes with the blue form to produce a green solution. As the solution approaches an acid concentration of 1.0 M (pH 0.0), the color is intense green. When the concentration approaches 4 M, the blue form has changed completely to the yellow form and the solution is now bright yellow.

▲ From the structure under Reactions, the H⁺ ion present in excess in the acid solution would likely attach to the negative site of the SO_3^-. This attachment would change the character of the structure and change its absorbance. The intermediate green color results from the combined effect of the yellow form and the blue form, both present in the test tube.

Reactions

FD&C Blue No. 1   FD&C Blue No. 1
(blue)   (yellow)

Notes

1. Be sure the food coloring contains FD&C Blue No. 1, which is the same as erioglaucine or C. I. Acid Blue No. 9.

2. FD&C Blue No. 1 is a suspected carcinogen, yet it is allowed as a food coloring. Possibly, the amount needed to cause cancer in laboratory animals is much larger than the amount that could potentially be consumed by humans. Also, there may be conflicting research.

3. The concentration of HCl and the corresponding colors of the FD&C Blue No. 1 are as follows:

[HCl] (M)	Color
6.0	yellow
5.0	yellow
4.0	yellowish-green
3.0	yellow–green
2.0	green
1.0	dark green
0.5	blue–green
0.1	blue

References

Lecture Demonstration in General Chemistry, 1st ed.; McGraw Hill: New York, 1939.

The Merck Index, 10th ed.; Merck: Rahway, NJ, 1983; p 250.

Witt, O. N. *Ber. Dtsch. Chem. Ges.* **1886**, *19*, 3121.

Materials

▲ FD&C Blue No. 1, blue food coloring, available in supermarkets.

▲ Hydrochloric acid, 6.0 M HCl: Add 49.6 mL of concentrated acid to enough water to make 100 mL of solution. Always add acid to water.

▲ Add 5.0 mL of the 6.0 M acid to 1.0 mL of distilled water to make the 5.0 M solution.

▲ Add 4.0 mL of the 6.0 M acid to 2.0 mL of distilled water to make the 4.0 M solution.

▲ Add 3.0 mL of the 6.0 M acid to 3.0 mL of distilled water to make the 3.0 M solution.

▲ Add 2.0 mL of the 6.0 M acid to 4.0 mL of distilled water to make the 2.0 M solution.

▲ Add 1.0 mL of the 6.0 M acid to 5.0 mL of distilled water to make the 1.0 M solution.

▲ Add 1.0 mL of the 6.0 M acid to 11.0 mL of distilled water to make the 0.5 M solution.

▲ Add 1.0 mL of the 6.0 M acid to 59.0 mL of distilled water to make the 0.1 M solution.

Oxidation–Reduction

A New Copper Mirror

A test tube is rinsed with nitric acid, tin(II) chloride, and silver nitrate solutions. A mixture of copper(II) sulfate and sodium potassium tartrate solutions is heated in the rinsed test tube with 40% glyoxal. A copper mirror develops on the sides of the test tube.

Procedure

Please consult the Safety Information before proceeding.

1. Use steel wool to thoroughly roughen the inside of a small test tube (see Notes).

2. Wash the inside of the test tube with several milliliters of concentrated nitric acid. Pour this solution out.

3. Rinse the tube with about 1.5 mL of acidified tin(II) chloride solution. Pour this solution out.

Safety Information

1. Concentrated nitric acid is corrosive and a strong oxidant, and the fumes are toxic by inhalation.

2. Tin(II) chloride (aqueous) is a toxic, corrosive solution. It is also a skin irritant and should be treated with care because the solution has been acidified.

3. Silver nitrate solution is toxic and corrosive. Avoid contact with the eyes and skin. →

4. Copper(II) sulfate (aqueous) may be harmful by inhalation, ingestion, or skin absorption.

5. Never use a solution of glyoxal that is stronger than commercial glyoxal, 40%. Glyoxal can be absorbed through the skin and is also flammable. It is moderately irritating to the skin and mucous membranes.

4. Rinse the tube with about 1.5 mL of silver nitrate solution. Pour out the solution and dispose of it properly according to the procedures in Appendix 3.

5. Mix 2.0 mL of copper(II) sulfate solution with 2.0 mL of basic sodium potassium tartrate solution in a small beaker.

6. Pour the sulfate–tartrate solution into the rinsed test tube.

7. Add 1.5 mL of 40% glyoxal to the test tube. Stir the contents to mix.

8. Place the test tube in a hot water bath until the solution starts to bubble. As a safety precaution, use a hot plate to heat the water bath. If you use a Bunsen burner, extinguish the flame before placing the test tube in the bath.

9. Remove the test tube with a holder and observe the Cu mirror that develops on the sides of the test tube.

Concepts

▲ This a reduction–oxidation reaction. Cu^{2+} is reduced to Cu^0, and glyoxal is oxidized, probably to glyoxylic acid.

▲ When a substance is reduced, the oxidation number changes from a higher number to a lower number. Electrons are gained in this process. The substance that is reduced is the oxidizing agent.

▲ When a substance is oxidized, the oxidation number changes from a lower number to a higher number. Electrons are lost in this process. The substance that is oxidized is the reducing agent.

▲ See the discussion of formalin versus glyoxal under Concepts of Investigation 50.

▲ One of the first descriptions of electroless plating was by Justus von Liebig in 1835, and was concerned with the reduction of silver salts by reducing aldehydes.

▲ The higher cost of chemical reducing substances compared with that of electricity prevented much progress being made in electroless plating until after World War II. Then the development of alloys, notably nickel–phosphorus, and printed circuits and the large-scale introduction of plastics resulted in the growth of electroless plating.

▲ In electroless plating, electrons are supplied by chemical reducing agents such as formaldehyde and hypophosphite as opposed to electrons being supplied by an external source such as a battery or generator in electroplating. The metals that are reduced are mostly nickel, copper, gold, and silver.

▲ Some of the modern uses for electroless plating include plating on plastics, especially for automotive items; printed circuits, used in communications, instruments, military, aerospace, and business applications; plating on glass such as mirror production, which uses electroless silver; the use of translucent metal films on commercial glass-plating for reduction of environmental heat gain; and ceramic resistors.

Materials

▲ Nitric acid, concentrated: commercial stock solution.

▲ Acidified tin(II) chloride, $SnCl_2$: Dissolve 1.0 g in 100 mL of 6 M HCl.

▲ Silver nitrate, $AgNO_3$: Dissolve 1.0 g in 100 mL of water.

▲ Copper(II) sulfate pentahydrate, $CuSO_4 \cdot 5H_2O$: Dissolve 13.9 g in 100 mL of water.

▲ Basic sodium potassium tartrate tetrahydrate, $KNaC_4H_4O_6 \cdot 4H_2O$: Dissolve 9.02 g of tartrate and 18.0 g of NaOH in 100 mL of water.

▲ Glyoxal, 40% CHOCHO: Purchase as 40% solution. A precipitate may form because of prolonged storage, but it can be redissolved by warming to 50–60 °C.

Reactions

Experimentation indicates there is a 1:1 mole ratio between the glyoxal used and the copper produced. Testing the final solution with 0.1 M Ca^{2+}(aq) did not produce a precipitate; this result indicates that oxalic acid is probably not a product. These findings suggest the reaction

$$CHOCHO(aq) + Cu^{2+}(aq) + 2OH^-(aq) \rightarrow$$
$$\text{glyoxal} \qquad Cu(s) + CHOCOOH(aq) + H_2O(l)$$
$$\text{glyoxylic acid}$$

Notes

1. Although it is somewhat difficult to do, it is important to thoroughly roughen the inside of the test tube. Twist a piece of steel wool so it fits inside the test tube, stuff it into the test tube, use forceps to hold onto it, and turn it a number of times. Make sure you wrap the test tube in a towel to protect your hand in case the test tube breaks.

2. If you teach more than one chemistry class a day, make up enough of the sulfate–tartrate solution to be used for all of your classes. It will last all day, but not longer.

References

Ealy, Julie B. *Sci. Teach.* **1991**, *58*(4), 26.

Hill, J. W.; Foss, D. L.; Scott, L. W. *J. Chem Educ.* **1979**, *56*, 752.

Lindner, V.; Krulick, G. A. In *Kirk–Othmer Encyclopedia of Chemical Technology,* 3rd ed.; Grayson, M.; Eckroth, D., Eds.; Wiley: New York, 1979; Vol. 8, pp 738–750.

Summerlin, L.; Borgford, C.; Ealy, Julie B., *Chemical Demonstrations: A Sourcebook for Teachers;* American Chemical Society: Washington, DC; 1987; Vol. 2, p 187.

A Reddish Gold Solid of Copper(I) Oxide

Solutions of copper(II) sulfate and glucose are mixed and heated to boiling. Upon addition of a solution of sodium hydroxide, a bright gold color works its way up through the solution, which produces reddish gold copper(I) oxide after filtration.

Procedure

Please consult the Safety Information before proceeding.

1. Place 40 mL of copper(II) sulfate solution and 150 mL of glucose solution in a 400-mL beaker. Bring the solution to a boil.

2. Add 35 mL of sodium hydroxide solution. Boil the solution 10 min, while observing the appearance of a bright gold slurry.

3. Cool the solution in an ice bath. After the solid settles, decant the alkaline solution and filter the slurry.

Safety Information

1. Copper sulfate pentahydrate is toxic if ingested and may be harmful if it comes in contact with mucous membranes or if the dust is inhaled.

2. Sodium hydroxide is corrosive to all tissues. Inhalation of the dust or concentrated mist may cause damage to the respiratory tract. Wear gloves, goggles, and a rubber apron when handling.

3. Although glucose is a food substance, no substance should ever be tasted in the laboratory. →

265

4. Concentrated ammonia, when inhaled, causes edema of the respiratory tract, spasm of the glottis, and asphyxia.

5. External contact with hydrochloric acid causes severe burns, and contact with the eyes may result in a total loss of vision. Inhalation causes coughing and choking with possible inflammation and ulcerations of the respiratory tract. Work in a fume hood, and wear goggles, gloves, and a rubber apron.

4. Wash the solid twice and show it to the class.

5. Place a pea-sized piece of solid on a watch glass. Add 3 drops of concentrated ammonia.

6. Repeat Step 5 with concentrated hydrochloric acid.

7. Repeat Step 5 with water.

Concepts

▲ This reaction illustrates reduction–oxidation. Cu^{2+} is reduced to Cu^+. Copper(II) gains an electron to form copper(I). Because Cu^{2+} is reduced, it acts as the oxidizing agent.

▲ Glucose acts as the reducing agent. It is oxidized and loses electrons.

▲ Copper(I) oxide is practically insoluble in water but soluble in ammonia and hydrochloric acid. Copper(II) oxide is practically insoluble in water, is soluble in hydrochloric acid, and is slowly soluble in ammonia.

▲ One of the earliest references to copper(I) oxide seems to be that of G. Agricola (1546), who referred to it as "red copper".

Reactions

1. Reduction:

$$2Cu^{2+}(aq) + 2OH^-(aq) + 2e^- \rightarrow Cu_2O(s) + H_2O(l)$$

2. Oxidation (glucose is represented by R–CHO):

$$R\text{–}CHO(aq) + 2OH^-(aq) \rightarrow$$
$$R\text{–}COOH(aq) + H_2O(l) + 2e^-$$

Notes

1. Having the three solutions ready before class allows you to do the demonstration in a reasonable amount of class time.

2. You might want to compare the appearance of copper(II) oxide with the synthesized copper(I) oxide after it dries.

3. Compare the solubility and reaction of copper(II) oxide in water, concentrated ammonia, and concentrated hydrochloric acid with that of copper(I) oxide.

References

Agricola, G. *Interpret. Germ. Metall. Basil.* **1546**, 462.

Mellor, J. W. *Comprehensive Inorganic Chemistry*; Pergamon: New York; 1973; p 117.

Stone, C. H. *J. Chem. Educ.* **1944**, *21*, 350.

Materials

▲ Copper(II) sulfate pentahydrate, $CuSO_4 \cdot 5H_2O$: Dissolve 10 g in 40 mL of water.

▲ Glucose, $C_6H_{12}O_6$: Dissolve 3 g in 150 mL of water.

▲ Sodium hydroxide, NaOH: Dissolve 7 g in 35 mL of water.

▲ Ammonia, concentrated NH_3: commercial stock solution.

▲ Hydrochloric acid, concentrated HCl: commercial stock solution.

Comparison of the Reducing Rates of Simple Sugars

Solutions of potassium hydroxide and glucose are mixed. Methylene blue indicator is added, and the solution turns blue. The blue color fades in a specific amount of time. A different amount of time for the blue color to fade results when galactose or fructose is used in place of glucose.

Procedure

Please consult the Safety Information before proceeding.

1. Place 100 mL of potassium hydroxide solution in a 250-mL flask.

2. Add 100 mL of glucose solution.

3. Add 1 mL of methylene blue indicator and swirl the solution to mix. Stopper the flask and begin timing. When the blue color has faded, record the time.

Safety Information

1. Potassium hydroxide is harmful if swallowed, inhaled, or absorbed through the skin. The material is extremely destructive to the tissue of the mucous membranes and the upper respiratory tract, eyes, and skin. Wear gloves when handling.

2. Although some simple sugars are food substances, no substance in the laboratory should ever be tasted.

3. Methylene blue may be harmful by inhalation, ingestion, or skin absorption. It causes severe eye irritation.

Materials

▲ Potassium hydroxide, KOH: Dissolve 26.6 g KOH in 500 mL of distilled water.

▲ Sugar, $C_6H_{12}O_6$: Dissolve 33.2 g of each sugar separately in enough distilled water to make 500 mL of solution. Order the α-D-simple sugars: glucose, galactose, and fructose.

▲ Methylene blue.

4. Repeat Steps 1–3 with the galactose and fructose solutions.

5. The flasks can be shaken more than once to obtain the blue color again.

Concepts

▲ Methylene blue, the indicator, is colorless in its reduced form. In the presence of oxygen, it is oxidized to the blue form.

▲ A blue color can be observed at the interface between the gas (the oxygen in the flask) and the solution. This blue color demonstrates that a gas is entering the solution.

▲ The initial bluing reaction is quite fast, but the disappearance of the blue color takes much longer. In a series of fast reactions, alkaline glucose is converted to glucoside, oxygen in the air in the flask goes into solution, and dissolved oxygen oxidizes the reduced colorless methylene blue to the blue oxidized form. In a slow reaction, the blue oxidized form is reduced by the glucoside, and it is not until this reaction is complete that a fully colorless solution can be observed.

▲ Glucose and galactose are stereoisomeric aldoses, and fructose is a ketose sugar. The structural formulas are

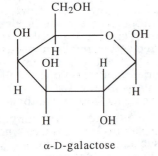

α-D-galactose

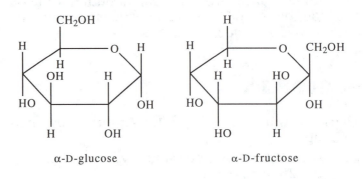

α-D-glucose   α-D-fructose

▲ A close relationship between the ability of a simple sugar to support the life of an organism and the ease with which it is oxidized seems to exist. Bertolf (1927) reared honeybee larvae on pure sugars and found that fructose, glucose, and galactose support life in the order given.

In studies done with the honeybee *Lucilla sericata*, Abbott (1947) found that most survive 33 days on fructose, 29 days on glucose, and only 14 days on galactose; he found that a solution made with dry powdered hemoglobin was reduced by 0.02 M fructose in 14 min, by glucose in 16 min, and by galactose in 20 min.

▲ Glucoside is present in the reaction. When methanol reacts with glucose in the presence of dilute acid, a mixture of acetals, the α- and β-methyl glucosides, are produced. Acetals of glucose are termed glucosides. The structure of methyl α-D-glucoside is

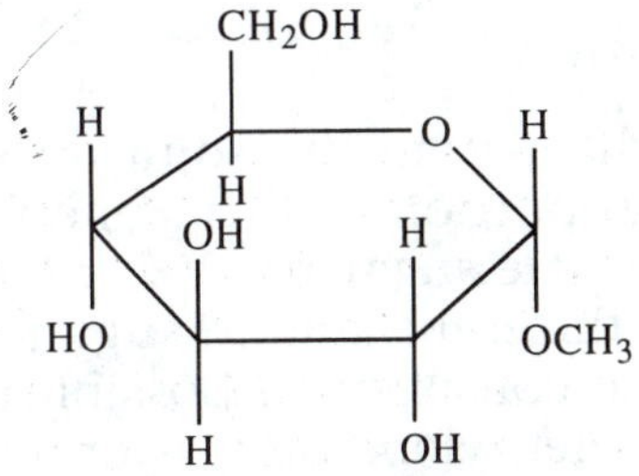

methyl α-D-glucoside

(Glucoside)⁻ is

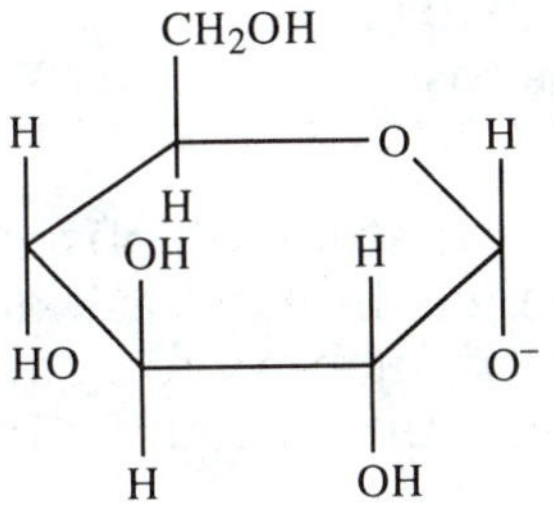

(glucoside)⁻

Reactions

In the following mechanism; G is glucose, G⁻ is glucoside, X is reduced colorless methylene blue, and B is oxidized blue methylene blue.

The four steps in the mechanism are

1. fast: $G + OH^- \leftrightarrows G^- + H_2O$
2. fast: $O_2(g) \rightarrow O_2(aq)$
3. fast: $O_2(aq) + X \rightarrow B$
4. slow: $B + G^- \rightarrow X + products$

   Net: $G + O_2(g) \rightarrow products$

The rate of the reaction is dependent on the concentration of sugar, hydroxide ion, and methylene blue but not on the concentration of the oxygen in the air.

Notes

1. The length of time the solution stays blue is directly proportional to the amount of shaking and the simple sugar used. For comparison of the three different simple sugars, it is important to be as consistent as possible with the mixing of the methylene blue after its addition.

2. Following are approximate times for the blue color to fade on the basis of the sugar used:

Sugar	Time (s)
Glucose	95
Galactose	85
Fructose	7

3. Make sure your students observe the interface between the gas in the flask and the solution. The color is light blue for galactose and glucose and light lime green for fructose.

References

Abbott, G. E. *Proc. Arkansas Acad. Sci.* **1947**, *2*, 45.

Abbott, G. E. *J. Chem. Educ.* **1948**, *26*, 100.

Campbell, F. B. *J. Chem. Educ.* **1963**, *40*, 578.

Kemp, D. S.; Vellaccio, F. *Organic Chemistry;* Worth: New York, 1980; p 983.

Six Oxidation States of Manganese

When one or more colorless solutions are added to a purple solution of potassium permanganate, it undergoes dramatic color changes.

Procedure

Please consult the Safety Information before proceeding. Where drops are indicated, use a microtip pipet.

1. Place 40 mL of potassium permanganate solution in each of five 150-mL beakers, labeled A, B, C, D, and E. Save beaker A to demonstrate the purple color of MnO_4^- ions in solution. The solution in beaker A represents manganese(VII).

2. Place 40 mL of manganese(II) chloride in a sixth 150-mL beaker. The light pink color represents Mn(II) ions in solution.

3. Beaker B, Mn(VI): Add 20 drops of 6 M NaOH and stir the solution to mix. Stirring after the

Safety Information

1. Manganese(II) chloride may be harmful by inhalation, ingestion, or skin absorption. It causes eye and skin irritation.

2. Hydrogen peroxide, 30%, is a strong oxidant and should be used with caution. Avoid all contact with the skin and eyes. Wear rubber gloves and goggles.

3. Dilute solutions of potassium permanganate are mildly irritating, and high concentrations are caustic. →

4. Sodium hydroxide is corrosive to all tissues. Inhalation of the dust or concentrated mist may cause damage to the respiratory tract. Wear gloves, goggles, and a rubber apron when handling.

5. Concentrated sulfuric acid is very corrosive. Handle with caution and avoid contact with the skin because it produces severe burns. Inhalation of the concentrated vapor may cause serious lung damage. Work in a fume hood, and wear gloves, goggles, and a rubber apron.

addition of each drop, add approximately 10 drops of 6% H_2O_2 to obtain a green color.

4. Beaker C, Mn(V): Add 36 mL of 50% NaOH and stir the solution to mix. Then, one by one with stirring after the addition of each dropperful, add approximately 5 dropperfuls of 50% NaOH until a dark blue color is obtained.

5. Beaker D, Mn(IV): Add 20 drops of 6 M NaOH and stir the solution to mix. Add approximately 20 drops of 6% H_2O_2, stirring after the addition of each drop, to obtain a brown precipitate.

6. Beaker E, Mn(III): Add 1 drop of concentrated H_2SO_4 and stir the solution to mix. Add approximately 3 drops of 6% H_2O_2, stirring after the addition of each drop, to obtain a rose color.

7. Display all the beakers, in a row, with decreasing oxidation states.

Concepts

▲ Hydrogen peroxide is a reducing agent with stronger oxidizing agents. Water and oxygen gas are often the products. The peroxide ion, O_2^{2-}, is oxidized to O_2^0 with the loss of electrons.

▲ Potassium permanganate is a strong oxidizing agent. Mn(VII) is reduced to form an ion with a lower oxidation number by gaining electrons.

▲ Manganese's outermost electron configuration is $3d^5 4s^2$. Manganese has a maximum of six different possible oxidation states. The $4s^2$ electrons are lost first and then the 3d electrons. Its most common oxidation states are II, IV, and VII.

▲ Louis-Jacques Thenard, because of his prompt publication, is credited with the discovery of hydrogen peroxide in 1818.

Reactions

1. Mn(VII) to Mn(VI):

$2MnO_4^-(aq) + 2OH^-(aq) + H_2O_2(aq) \rightarrow$
purple $2MnO_4^{2-}(aq) + O_2(g) + 2H_2O(l)$
green

2. Mn(VII) to Mn(V):

$$2MnO_4^-(aq) + 4OH^-(aq) \rightarrow$$
purple $\qquad 2MnO_4^{3-}(aq) + O_2(g) + 2H_2O(l)$
blue

3. Mn(VII) to Mn(IV):

$$2MnO_4^-(aq) + 3H_2O_2(aq) \rightarrow$$
purple
$$2MnO_2(s) + 3O_2(g) + 2H_2O(l) + 2OH^-(aq)$$
brown

4. Mn(VII) to Mn(III):

$$MnO_4^-(aq) + 4H^+(aq) + 2H_2O_2(aq) \rightarrow$$
purple $\qquad Mn^{3+}(aq) + 2O_2(g) + 4H_2O(l)$
rose

Materials

▲ Potassium permanganate, $KMnO_4$: Dissolve 0.22 g in 400 mL of water.

▲ Manganese(II) chloride, $MnCl_2$: Dissolve 10 g in 100 mL of distilled water.

▲ Hydrogen peroxide, 6% H_2O_2: Mix 20 mL of 30% with 80 mL of distilled water.

▲ Sodium hydroxide, 6 M NaOH: Dissolve 12 g in enough water to make 50 mL of solution.

▲ Sodium hydroxide, 50% NaOH: Dissolve 50 g in 50 mL of distilled water.

▲ Sulfuric acid, concentrated H_2SO_4: commercial stock solution.

Notes

1. Any manganese compound in which manganese has a 2+ oxidation state could be used to make up a solution representing the pink color of the ion.

2. A 9% solution is the usual concentration of hydrogen peroxide used for hair dyeing.

References

Arora, C. L. *J. Chem. Educ.* **1977**, *54*, 302.

Cotton, F. A.; Wilkinson, G. *Advanced Inorganic Chemistry*, 4th ed.; Wiley: New York, 1980; pp 738–749.

Kolb, D. *J. Chem. Educ.* **1988**, *65*, 1004.

Nebergall, W. H.; Schmidt, F. C.; Holtzclaw, H. F., Jr. *College Chemistry*, 5th ed.; Heath: Lexington, MA, 1976; pp 264, 896.

Schumb, W. C.; Satterfield, C. N.; Wentworth, R. L. *Hydrogen Peroxide;* Reinhold: New York, 1955; pp 1–2.

From Wine to Water and Water to Wine

A colorless acidified sodium hydrogen sulfite solution is added to a red potassium iodide–iodine solution, and the mixture becomes colorless. A colorless potassium thiocyanate solution is added to a beaker containing a few crystals of iron(III) nitrate nonahydrate, and the solution becomes dark red.

Procedure

Please consult the Safety Information before proceeding.

1. Place 150 mL of red KI-I$_2$ solution in a 250-mL beaker. Place the beaker on a magnetic stirrer and turn on at low speed.

2. When ready, pour 60 mL of acidified sodium bisulfite into the KI-I$_2$ solution, leaving the zinc behind. The solution will become colorless or nearly so.

3. Place a few crushed crystals of iron(III) nitrate

5. Potassium thiocyanate may be harmful by inhalation, ingestion, or skin absorption. It causes eye and skin irritation. It is irritating to mucous membranes and the upper respiratory tract.

6. External contact with hydrochloric acid causes severe burns, and contact with the eyes may result in a total loss of vision. Inhalation causes coughing and choking with possible inflammation and ulceration of the respiratory tract. Work in a fume hood, and wear goggles, gloves, and a rubber apron.

nonahydrate in a 250-mL beaker on a magnetic stirrer.

4. When ready, pour 200 mL of the potassium thiocyanate solution into the beaker containing the crystals. Turn on the stirrer. A dark red solution results.

Concepts

▲ Iodine is not very soluble in water, but when combined with a solution of potassium iodide, a red solution is produced. This color change is the result of the formation of I_3^-, which is the red triiodide complex between iodine and the iodide ion. Iodine is kept in solution by this complex.

▲ Zinc metal acts as a reducing agent, losing electrons and forming $Zn^{2+}(aq)$. The electrons are gained by the hydrogen sulfite ion, HSO_3^-, forming the thiosulfate ion, $S_2O_3^{2-}$.

▲ The thiosulfate ion then acts as a reducing agent, losing electrons and forming the tetrathionate ion, $S_4O_6^{2-}$. The electrons are gained by I_3^-, forming colorless I^- ions.

▲ When iron(III) ions, Fe^{3+}, combine with thiocyanate ions, SCN^-, the dark red iron(III) thiocyanate complex is formed.

Reactions

1. $I^-(aq) + I_2(s) \rightarrow I_3^-(aq)$
   colorless   red

2. $2HSO_3^-(aq) + 2Zn(s) + 4H^+(aq) \rightarrow$
   $\qquad\qquad\qquad 2Zn^{2+}(aq) + S_2O_3^{2-}(aq) + 3H_2O(l)$

3. $2S_2O_3^{2-}(aq) + I_3^-(aq) \rightarrow S_4O_6^{2-}(aq) + 3I^-(aq)$
   $\qquad\qquad\qquad$ red $\qquad\qquad\qquad\qquad\qquad$ colorless

4. $Fe^{3+}(aq) + SCN^-(aq) \rightarrow FeSCN^{2+}(aq)$
   orange $\qquad$ colorless $\qquad$ dark red

Notes

1. If the sodium bisulfite solution alone is used without the acid and granular zinc, the result-

ing solution is light yellow. You might want to compare this reaction with the demonstration as written.

2. A few crystals of sodium thiosulfate dissolved in 150 mL of water could be used in place of the NaHSO$_3$–Zn–HCl solution.

References

DuBois, R. *J. Chem. Educ.* **1937**, *14*, 324–326.

Greenwood, N. N.; Earnshaw, A. *Chemistry of the Elements;* Pergamon: New York, 1984; pp 847, 852–853, 1263.

Kohn, M. *J. Chem. Educ.* **1943**, *20*, 198.

Lippy, J. D., Jr.; Palder, E. L. *Modern Chemical Magic;* Stackpole: Harrisburg, PA, 1959; p 2.

Manahan, S. E. *Quantitative Chemical Analysis;* Brooks/ Cole: Monterey, CA, 1986; pp 348, 351–352.

Materials

▲ Potassium iodide–iodine, KI–I$_2$: Dissolve 30 g of potassium iodide in 150 mL of water. This mixture will have a slightly yellow tinge. Adding a few crystals of iodine at a time, stir the mixture until the crystals are dissolved. Continue to add iodine, a few crystals at a time, until a red color is obtained.

▲ Acidified sodium hydrogen sulfite, Na$_2$SO$_3$: Dissolve 7.8 g in 60 mL of water. Add 2 g of 20-mesh (granular) zinc and 3 drops of concentrated hydrochloric acid. Stir the mixture for several minutes on a magnetic stirrer before using it.

▲ Potassium thiocyanate, 0.02 M KSCN: Dissolve 1.9 g in enough water to make 1 L of solution.

▲ Iron(II) nitrate nonahydrate, crystals.

Nitrite Ion Oxidizes and Reduces

Upon addition of an acidified solution of potassium nitrite, a red solution of basic fuchsin or a purple solution of potassium permanganate goes through several color changes until becoming colorless.

Procedure

Please consult the Safety Information before proceeding.

1. Place 40 mL of acidified potassium nitrite solution in a 100-mL beaker.

2. Add 5 mL of potassium permanganate solution.

3. Swirl to mix and observe color changes from purple to yellow to colorless.

4. Repeat Steps 1–3 using basic fuchsin solution instead of potassium permanganate, observing color changes from red to purple to blue to turquoise to colorless.

4. Sulfuric acid is very corrosive. Handle with caution and avoid contact with the skin since it produces severe burns. Inhalation of the concentrated vapor may cause serious lung damage. Work in a fume hood, wear goggles, gloves, and a rubber apron.

Concepts

▲ As a reducing agent, acidified NO_2^- is oxidized to NO_3^-. Electrons are lost as the 3+ oxidation number of the nitrogen in the nitrite ion increases to 5+ in the nitrate ion. MnO_4^- is the oxidizing agent as Mn(VII) is reduced to Mn(II) by gaining electrons.

▲ As an oxidizing agent, acidified NO_2^- is reduced to NO. Electrons are gained as the 3+ oxidation number of nitrogen in the nitrite ion decreases to 2+ in nitric oxide, NO. Basic fuchsin is the reducing agent since it is oxidized by losing electrons.

▲ Elements that have at least three oxidation states are capable of undergoing disproportionation. Common examples are N, P, S, Cl, Br, I, Mn, Cu, Au, and Hg. For disproportionation to happen, the element in the reactant must have one higher and one lower oxidation state. This condition is true for nitrogen in the nitrite ion and the nitrogen products, NO and NO_3^-, that it forms in the two separate reactions.

▲ The transmitted colors of the oxidation of basic fuchsin—red → purple → blue → bluish green—result from a delocalized electronic charge associated with the color-bearing part of the molecule.

▲ The chemical name of fuchsin is 3-methyl-4,4'-diaminofuchsonimonium chloride. The structural formulas in its colored form and colorless, or carbinol base, form are shown under Reactions.

▲ Basic fuchsin is an example of a dye in a series called triphenylmethanes. These dyes were among the earliest classes of synthetic organic dyes to achieve commercial success. The triphenylmethane dyes are mainly derivatives of triphenylmethane, $(C_6H_5)_3CH$, and diphenylnaphthylmethane, $(C_6H_5)_2CH(C_{10}H_7)$. Basic fuchsin was discovered by Natanson (1856).

▲ Most triphenylmethane dyes are destroyed under strong oxidizing conditions. Primary amino-substituted triphenylmethanes are especially vulnerable to oxidation. Sodium hypochlorite bleach or prolonged sunlight exposure in air destroys the color of the dyes.

▲ The redox properties of the dyes that permit color generation and reversal to colorless dye precursors are important in the photographic and duplicating fields. Photolytic or electrical stimuli activate oxidants or reductants that affect the dyes. Under stringent manufacturing controls and special purification procedures, some of the dyes are used as food colors.

▲ The basic principles illustrated in this demonstration are oxidation-reduction, disproportionation, and dyes as reducing agents.

Materials

▲ Basic fuchsin: Dissolve 0.05 g in 100 mL of distilled water.

▲ Potassium permanganate, $KMnO_4$: Dissolve 0.08 g in 100 mL of distilled water.

▲ Acidified potassium nitrite, 1.0 M KNO_2: Dissolve 8.5 g in enough water to make 100 mL of solution. Add one drop of concentrated sulfuric acid.

Reactions

1. Disproportionation:

$$3HNO_2(aq) \rightarrow 2NO(g) + NO_3^-(aq) + H_3O^+(aq)$$

2. Nitrite ion as a reducing agent:

$$5NO_2^-(aq) + 6H^+(aq) + 2\underset{\text{purple}}{MnO_4^-}(aq) \rightarrow$$

$$2Mn^{2+}(aq) + 5NO_3^-(aq) + 3H_2O(l)$$
$$\underset{\text{colorless}}{\phantom{2Mn^{2+}(aq)}}$$

3. Colored and colorless forms of basic fuchsin:

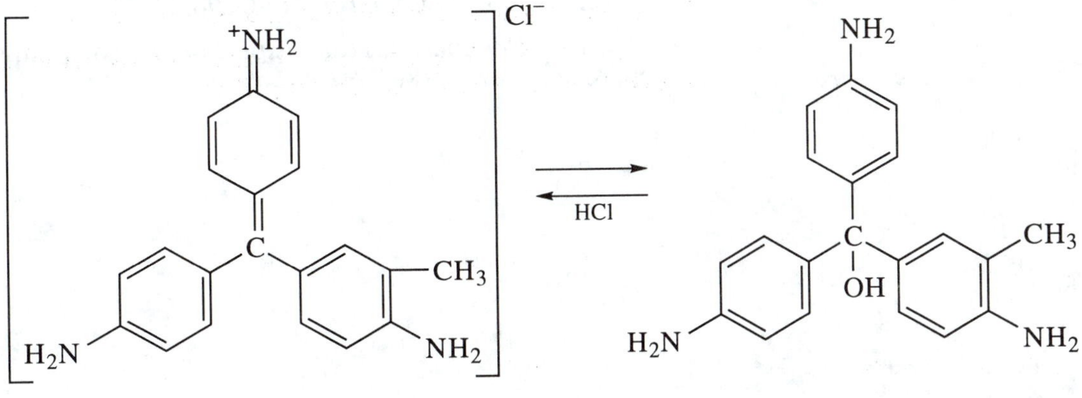

basic fuchsin
(colored)

carbinol
(colorless)

Notes

1.   The reduction probably stops at Mn^{2+}. These ions are light pink in solution. The solution usually looks colorless, though, possibly because of the low concentration of Mn(II).

2.   You might want to use an overhead projector so the color changes can be observed more easily. Otherwise, your students should be close to the beakers.

3.   You can continue to add 5-mL portions of $KMnO_4$ or basic fuchsin solution to the KNO_2 solution an additional four or five times and continue to observe the color changes.

References

Bailar, J. C., Jr.; Moeller, T.; Kleinberg, J.; Guss, C. O.; Castellion, M. E.; Metz, C. *Chemistry;* Academic: Orlando, Florida, 1984; pp 543–544.

Colour Index, 3rd ed.; Society of Dyers & Colourists, 1971; Vol. 4, p 4389.

Lindner, V.; Bannister, D.; Elliott, J. In *Kirk–Othmer Encyclopedia of Chemical Technology,* 3rd ed; Grayson, M.; Eckroth, D., Eds.; Wiley-Interscience: New York, 1983; Vol. 23, pp 399–412.

Natanson. *Ann. Chem. Pharm.* **1856,** *98,* 297.

Newth, G. S. *Chemical Lecture Experiments;* Longmans, Green, & Co.: New York, 1892; p 139.

Oxidation of Various Metal Hydroxides

Adding colorless potassium hydroxide solution to solutions of cobalt(II) nitrate, iron(II) sulfate, and manganese(II) sulfate results in the formation of blue, green, and yellow precipitates, respectively. Subsequent addition of hydrogen peroxide forms brown, orange, and black precipitates, respectively.

Procedure

Please consult the Safety Information before proceeding.

1. Place 10 mL of the cobalt(II) nitrate solution in a test tube.

2. Add 10 mL of the potassium hydroxide solution. **Do not mix.**

3. Repeat Steps 1 and 2 using the iron(II) sulfate solution in place of the cobalt solution.

Safety Information

1. Wear gloves when handling potassium hydroxide. It is harmful if swallowed, inhaled, or absorbed through the skin. The material is extremely destructive to the tissue of the mucous membranes and the upper respiratory tract, eyes, and skin.

2. Cobalt(II) nitrate hexahydrate is an eye, skin, and mucous membrane irritant. It is moderately toxic by ingestion. →

3. Iron(II) sulfate hepta-hydrate may be harmful by inhalation, ingestion, or skin absorption. It causes eye and skin irritation. It is irritating to mucous membranes and the upper respiratory tract.

4. Manganese(II) sulfate monohydrate may be a tissue irritant.

5. Hydrogen peroxide, 3%, is used as an antiseptic; however, it may cause irritation to the mucous membranes and serious damage to the eyes.

Materials

▲ Cobalt(II) nitrate hexahydrate, 0.1 M $Co(NO_3)_2 \cdot 6H_2O$: Dissolve 2.9 g in enough water to make 100 mL of solution.

▲ Iron(II) sulfate heptahydrate, 0.1 M $FeSO_4 \cdot 7H_2O$: Dissolve 2.7 g in enough water to make 100 mL of solution.

▲ Manganese(II) sulfate monohydrate, 0.1 M $MnSO_4 \cdot H_2O$: Dissolve 1.7 g in enough water to make 100 mL of solution.

▲ Potassium hydroxide, 0.1 M KOH: Dissolve 5.6 g in enough water to make 100 mL of solution.

▲ Hydrogen peroxide, 3% H_2O_2: Buy as 3% from the drugstore or dilute 10 mL of 30% with 90 mL of water.

4. Repeat Steps 1 and 2 using the manganese(II) sulfate solution in place of the cobalt solution.

5. Place the three test tubes in a test-tube rack. Have students observe the colors of the three precipitates.

6. Using a pipet, add 1 mL of the hydrogen peroxide solution to each of the test tubes. **Do not mix**. Have students observe the test tubes.

Concepts

▲ Following is a listing of the initial color of the compounds in solution and their precipitates with addition of base and then with peroxide:

Compound	Initial	With Base	With Peroxide
$Co(NO_3)_2$	red	blue	brown–black
$FeSO_4$	yellow	blue-green	orange-brown
$MnSO_4$	colorless	yellow	black

▲ Transition metal compounds often are colored. Once the compound is formed, the d orbitals in any one energy level of the metal are of different energies. Electronic transitions between orbitals occur with the absorption of visible light. The light that is transmitted or reflected is the complement of the absorbed light. For instance, if light corresponding to a wavelength of blue is absorbed, yellow light is transmitted.

▲ With addition of base, gel-like precipitates of cobalt(II), iron(II), and manganese(II) hydroxide are formed.

▲ Hydrogen peroxide acts as an oxidizing agent, and O_2^{2-} is reduced to O^{2-} by gaining electrons. Water is the product when hydrogen peroxide acts as an oxidizing agent.

▲ Each metal ion of the hydroxide is oxidized to a higher oxidation state by losing electrons and acts as a reducing agent.

▲ Transition metals can exhibit more than one oxidation state. The maximum state is given by the metal's group number, which is not usually the most stable oxidation state. For 3d transi-

tion elements, the most common oxidation
states are II, III, and IV.

▲   The metal ion also acts as a catalyst to produce
    oxygen gas from hydrogen peroxide. The de-
    composition of hydrogen peroxide is acceler-
    ated in the presence of heavy metals, easily
    oxidizable substances, and some other materi-
    als. Commercial solutions of this substance are
    stabilized by additives that are present to pro-
    tect against decomposition.

▲   The d transition metals and their compounds
    have high catalytic activity because of the ease
    with which their electrons are lost and gained
    or moved from one energy level to another.
    Examples in which this is important include the
    use of Fe_2O_3 in the Haber process to produce
    ammonia and the use of V_2O_5 for the produc-
    tion of sulfuric acid.

Reactions

$$Co^{2+}(aq) + 2OH^-(aq) \leftrightarrows Co(OH)_2(s)$$
  red   blue

$$Fe^{2+}(aq) + 2OH^-(aq) \leftrightarrows Fe(OH)_2(s)$$
  yellow   blue-green

$$Mn^{2+}(aq) + 2OH^-(aq) \leftrightarrows Mn(OH)_2(s)$$
  colorless   yellow

$$2Co(OH)_2(s) + H_2O_2(aq) \rightarrow 2CoO(OH)(s) + 2H_2O(l)$$
                                    brown-black

$$2Fe(OH)_2(s) + H_2O_2(aq) \rightarrow 2FeO(OH)(s) + 2H_2O(l)$$
                                    orange-brown

$$2Mn(OH)_2(s) + H_2O_2(aq) \rightarrow$$
$$2MnO(OH)(s) + 2H_2O(l)$$
                                    black

$$\overset{M^{2+}}{2H_2O_2(aq) \rightarrow O_2(g) + 2H_2O(l)}$$

where $M^{2+} = Co^{2+}$, Fe^{2+}, or Mn^{2+}

Notes

1. Sodium hydroxide, 0.1 M, could be used in place
    of potassium hydroxide.

2. Although the solutions could be mixed after addition of the sodium hydroxide and then the hydrogen peroxide, the observations are much more interesting if the solutions are not mixed. The students will be able to see the original precipitate on the bottom of the test tube and the new precipitate on the top after addition of hydrogen peroxide.

References

Bailar, J. C., Jr.; Moeller, T.; Kleinberg, J.; Guss, C. O.; Castellion, M. E.; Metz, C. *Chemistry*, 2nd ed.; Academic: Orlando, FL, 1984; p 1014.

Baum, B. M.; Finley, J. H.; Blumbergs, J. H.; Elliott, E. J.; Scholer, F.; Wouten, H. L. In *Kirk–Othmer Encyclopedia of Chemical Technology;* Wiley: New York, 1978; Vol. 3, p 938.

Fowles, G. *Lecture Experiments in Chemistry;* Blakiston: Philadelphia, PA, 1937; pp 251–252.

Greenwood, N. N.; Earnshaw, A. *Chemistry of the Elements;* Pergamon: New York, 1984; pp 1228 (manganese), 1302 (cobalt).

Nebergall, W. H.; Schmidt, F. C.; Holtzclaw, H. F., Jr. *College Chemistry;* Heath: Lexington, MA, 1976; pp 264–265.

Whitten, K. W.; Gailey, K. D. *General Chemistry with Qualitative Analysis*, 2nd ed.; Saunders: Philadelphia, PA, 1984; pp 779–782, 790.

Let's Keep It Iron(II)

Three test tubes containing solutions are left in the air for observations. An orange solution of iron(III) ammonium sulfate stays orange. A green solution of iron(II) ammonium sulfate turns yellow–orange. Another test tube containing the green solution of iron(II) ammonium sulfate and also iron filings stays green.

Procedure

Please consult the Safety Information before proceeding.

1. In a test tube, dissolve 1 g (1/2 tsp) of iron(III) ammonium sulfate dodecahydrate in 10 mL of water.

2. In a second test tube, dissolve 1 g (1/2 tsp) of iron(II) ammonium sulfate hexahydrate in 10 mL of water.

3. In a third test tube, dissolve 1 g (1/2 tsp) of

Safety Information

1. Iron(III) ammonium sulfate dodecahydrate is an eye, mucous membrane, and skin irritant.

2. Iron(II) ammonium sulfate hexahydrate is moderately toxic by ingestion and is an irritant through inhalation and eye and skin contact.

Materials

▲ To clean iron filings, submerge iron in ethanol, swirl the contents, and decant the black liquid. Rinse the filings two more times in ethanol. Rinse once in acetone. Air dry the filings. Avoid any open flames when using alcohol or acetone.

▲ Iron(II) ammonium sulfate hexahydrate, $Fe(NH_4)_2(SO_4)_2 \cdot 6H_2O$, powder.

▲ Iron(III) ammonium sulfate dodecahydrate, $FeNH_4(SO_4)_2 \cdot 12H_2O$, powder.

iron(II) ammonium sulfate hexahydrate in 10 mL of water. Add 0.5 g (1/4 tsp) of cleaned iron filings.

4. Leave the test tubes out for observations during the class period and for several days.

Concepts

▲ Iron(III) ions , Fe^{3+}, are stable in the air, and the color of the solution with iron(III) ammonium sulfate remains orange.

▲ Iron(II) ions, Fe^{2+}, are not stable in the air and undergo oxidation to form Fe^{3+}. The color of the iron(II) ammonium sulfate solution changes from light green to yellow–orange.

▲ In the test tube containing the iron filings, although iron(II) ions are oxidized to iron(III) ions, they are reduced again to Fe^{2+} by gaining six electrons from the iron filings. The iron filings or Fe^0 is then oxidized to Fe^{2+} by losing six electrons. The color remains light green even after several days.

Reactions

1. Oxidation of Fe^{2+}:

$$Fe^{2+}(aq) \xrightarrow{O_2} Fe^{3+}(aq) + e^-$$

$$3Fe^{2+}(aq) \xrightarrow{O_2} 2Fe^{3+}(aq)$$

2. Test tube containing the iron filings:

$$2Fe^{3+}(aq) + Fe^0(s) \rightarrow 3Fe^{2+}(aq)$$

References

Nebergall, W. H.; Schmidt, F. C.; Holtzclaw, H. F., Jr. *College Chemistry;* Heath: Lexington, MA, 1976; pp 907–908.

Synthesis and Reduction of Sodium Manganate

Heating sodium hydroxide and manganese(IV) oxide together produces a dark green solid, sodium manganate. Addition of concentrated sulfuric acid and then hydrogen peroxide changes the color of the solution from dark green to pink.

Procedure

Please consult the Safety Information before proceeding.

1. Dissolve 0.6 g of solid sodium manganate in 300 mL of water.

2. Add 9 drops of concentrated sulfuric acid. Stir the solution to mix.

3. Add 7 drops of hydrogen peroxide, with stirring after the addition of each drop, to obtain a muddy pink color.

4. Filter some of the solution through filter paper into a test tube. A light pink solution results.

4. Concentrated sulfuric acid is very corrosive. Handle with caution and avoid contact with the skin because it produces severe burns. Inhalation of the concentrated vapor may cause serious lung damage. Work in a fume hood and wear gloves, goggles, and an apron.

Concepts

▲   Heating solid sodium hydroxide and manganese(IV) oxide produces sodium manganate, Na_2MnO_4. Water is one of the products. This result is obvious near the beginning of heating when the bottom of the crucible looks wet and some of the solid seems to stick to the bottom of the crucible.

▲   After the addition of sulfuric acid, a purple color results, which is the oxidation of Mn(VI) in MnO_4^{2-} to Mn(VII) in MnO_4^-.

▲   Continued addition of sulfuric acid results in a brown solid settling out, which is manganese(IV) oxide, MnO_2. The oxidation number in this solid is 4+.

▲   When an element in one oxidation state is both oxidized and reduced, such as Mn(VI) being oxidized to Mn(VII) and reduced to Mn(IV), this is known as *disproportionation*. The element must be capable of having at least three oxidation states. Other elements that can be involved in disproportionation reactions are sulfur, copper, nitrogen, and the halogens.

▲   In an acidic environment and with the addition of hydrogen peroxide, Mn(IV) in MnO_2 is reduced to Mn(II). Mn(II) in solution is pink. This is evident when the solution is filtered and the filtrate is pink. Brown manganese(IV) oxide solid collects on the filter paper.

Reactions

1.   Solid sodium manganate:

$$4NaOH(s) + 2MnO_2(s) + O_2(g) \rightarrow$$
$$2Na_2MnO_4(s) + 2H_2O(l)$$
$$\text{green}$$

2. $3MnO_4^{2-}(aq) + 2H^+(aq) \rightarrow$
$$\text{green}$$

$$2MnO_4^-(aq) + 2OH^-(aq) + MnO_2(s)$$
$$\text{purple} \qquad\qquad\qquad \text{brown}$$

3. $MnO_2(s) + 2H^+ + H_2O_2 \rightarrow$
$$\text{brown}$$

$$Mn^{2+}(aq) + O_2(g) + 2H_2O(l)$$
$$\text{pink}$$

Notes

A beaker should be set up with 0.6 g of solid sodium manganate dissolved in 300 mL of water and left sitting out. Disproportionation or autoxidation-reduction of manganate ions to manganese(IV) oxide and permanganate ions will occur over several days. Students can record the changes that occur during this time. The reaction is

$$3MnO_4^{2-}(aq) + 2H_2O(l) \rightarrow$$
$$2MnO_4^-(aq) + 4OH^-(aq) + MnO_2(s)$$

References

Bailar, J. C., Jr.; Moeller, T.; Kleinberg, J.; Guss, C. O.; Castellion, M. E.; Metz, C. *Chemistry*, 2nd ed.; Academic: Orlando, FL, 1984; pp 543–544.

Greenwood, N. N.; Earnshaw, A. *Chemistry of the Elements;* Pergamon: New York, 1984; pp 1218–1219.

Lippy, J. D., Jr.; Palder, E. L. *Modern Chemical Magic;* Stackpole: Harrisburg, PA, 1959; p 15.

Nebergall, W. H.; Schmidt, F. C.; Holtzclaw, H. F., Jr. *College Chemistry*, 5th ed.; Heath: Lexington, MA, 1976; p 897.

Materials

▲ Hydrogen peroxide, 3% H_2O_2: Purchase from the drugstore.

▲ Sulfuric acid, concentrated H_2SO_4: commercial solution.

▲ Solid sodium manganate, Na_2MnO_4: In a fume hood, grind together 5 g of sodium hydroxide and 5 g of manganese(IV) oxide. The fume hood is especially important because of sodium hydroxide dust and possible pulmonary system damage from manganese(IV) oxide. Heat the mixture strongly in a crucible on a hot plate or with a Bunsen burner for 30 min. After 15 min of heating, remove the crucible from the hot plate and grind up any chunks. Continue heating the mixture for 15 more minutes.

Ring Opening of Phenolphthalein by Zinc Metal

When a pink solution of alkaline phenolphthalein indicator is added to zinc dust and heated, the indicator loses its color. When a few drops of acid are added to the decolorized solution, the color reappears.

Procedure

Please consult the Safety Information before proceeding.

1. Place 5 mL of distilled water in a large test tube. Add 1 g of zinc dust, 10 drops of phenolphthalein indicator, and 1 mL of sodium hydroxide solution. Swirl the contents to mix.

2. Heat the solution carefully over a Bunsen burner flame. Do not let the contents boil over. Observe the decolorization, which occurs very quickly, and remove the tube from the flame as soon as the solution is colorless.

Safety Information

1. Sodium hydroxide is corrosive to all tissues. Inhalation of the dust or concentrated mist may cause damage to respiratory tract. Wear gloves, an apron, and goggles when handling.

2. Concentrated hydrochloric acid causes severe burns, and contact with the eyes may result in a total loss of vision. Inhalation of the vapors may cause coughing and choking. Inflammation and ulceration of the respiratory tract may occur. →

3. Hydrogen peroxide, 30%, is a strong oxidant and should be used with caution. Avoid all contact with the skin and eyes; wear rubber gloves and goggles.

4. Ethanol is a volatile, highly flammable liquid. It should not be used near any open flames. Prolonged topical use may cause dryness, and inhalation may produce headaches.

3. Decant the solution into a clean test tube. Dropwise add 0.1 M hydrochloric acid until a slight pink color is restored to the solution.

4. Repeat Steps 1 and 2 and then decant the colorless solution into a test tube. Dropwise add 9% hydrogen peroxide until a slight pink color is restored.

Concepts

▲ When zinc metal is added to an alkaline solution of phenolphthalein, the zinc readily reduces the colored form of indicator to the colorless leuco form. The phenolphthalein oxidizes the zinc to 2+ in the zincate ion. Electrons are added to the oxygen in the para position on the ring. Double bonds are broken, and the color is lost.

▲ By adding hydrochloric acid, the leuco form is reduced back to the lactone form, a conjugated system is restored, and the color returns.

▲ Adding hydrogen peroxide also converts the leuco form back to the colored form. Shaking the solution in air will also accomplish the same reaction but at a much slower rate.

Reactions

lactone form (purple)   leuco form (colorless)

Notes

The H_2O_2 reaction is slow, and a tiny crystal of KI (catalyst) can be added to hasten it. The acid reaction with the leuco form is not completely understood.

References

Feigl, F. *Spot Tests, Vol. I, Inorganic Chemistry*, 4th English translation; Elsevier: Houston, TX, 1954; pp 273, 327.

Materials

▲ Zinc dust.

▲ Phenolphthalein indicator, 0.1%: Add 0.1 g to enough 95% ethanol to make 100 mL of solution.

▲ Hydrochloric acid, 0.1 M HCl: Add 0.9 mL of concentrated HCl to enough water to make 100 mL of solution.

▲ Hydrogen peroxide, 9%, H_2O_2: Use as purchased from a hair salon or add enough water to 30 mL of 30% H_2O_2 to make 100 mL of solution.

▲ Sodium hydroxide, 0.1 M NaOH: Add 0.4 g of solid NaOH to enough water to make 100 mL of solution.

Electrochemistry

Anodic and Cathodic Reactions

When the color changes involved with different indicators are used, the anodic and cathodic reactions of a piece of aluminum can be observed.

Procedure

Please consult the Safety Information before proceeding.

1. Gently boil the agar solution for 1 h in a hot water bath.

2. In four separate 100-mL beakers, preparing one at a time, mix 50 mL of agar solution, 6.5 mL of aluminon solution, and 7.5 mL of one of the indicators. Adjust the initial color of each solution as follows: turmeric, none; universal, 1 drop of 0.1 M NaOH to dark green; thymol blue, 1 drop of 0.1 M NaOH to greenish yellow; metacresol purple, 2 drops of 0.1 M HCl to orange.

6. Sodium hydroxide is corrosive to all tissues. Inhalation of the dust or concentrated mist may cause damage to the respiratory tract. Wear gloves, goggles, and a rubber apron when handling.

7. External contact with hydrochloric acid causes severe burns, and contact with the eyes may result in a total loss of vision. Inhalation causes coughing and choking with possible inflammation and ulceration of the respiratory tract. Work in a fume hood and wear goggles, gloves, and a rubber apron.

3. Pour enough of the solution into a Petri dish so that it is about half full. Let it cool to just hardening (over ice will help). Place the Al–Cu wire on the agar.

4. Cover the Al–Cu wire with as much remaining agar solution as needed.

5. Color changes will occur almost immediately and become progressively more obvious over an hour.

6. Store each plate on a wet piece of filter paper in a sealed plastic bag.

7. Make observations over several days.

Concepts

▲ Many metals undergo corrosion (oxidation) when exposed to air and water. When a metal is in contact with a less active metal, corrosion occurs faster.

▲ A metal such as Al, in contact with a less active metal, such as Cu, both in contact with moist air, make up an electrolytic cell.

▲ The surface of the Al wire consists of a series of tiny voltaic cells. Electrons are transferred through the wire, and the electrical circuit is completed by the flow of ions through the moist agar.

▲ The ends of the Al wire have been subjected to a strain because of cutting, and the oxidation of Al occurs there according to the anodic half-reaction:

$$Al(s) \rightarrow Al^{3+}(aq) + 3e^-$$

Reduction occurs along the central area in the vicinity of the Cu wire according to the cathodic half-reaction:

$$O_2(g) + 2H_2O(l) + 4e^- \rightarrow 4OH^-(aq)$$

A representative drawing of these reactions occurring with the Cu wire wrapped around the Al wire is as follows:

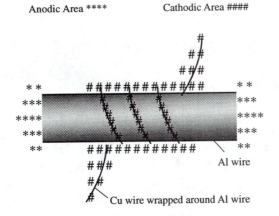

Anodic Area ****   Cathodic Area ####

Al wire

Cu wire wrapped around Al wire

Materials

▲ Agar: Mix 2.0 g of NaCl, 2.6 g of agar, and 200 mL of distilled water.

▲ Aluminon: Dissolve 0.1 g in 100 mL of distilled water.

▲ Sodium hydroxide, 0.02 M NaOH: Dissolve 0.08 g of NaOH in enough distilled water to make 100 mL of solution.

▲ Turmeric: Mix 0.1 g in 10 mL of 50% ethanol.

▲ Metacresol purple (alkaline range): Dissolve 0.1 g in 13.1 mL of 0.02 M NaOH, diluted to 250 mL with distilled water.

▲ Thymol blue (alkaline range): Dissolve 0.1 g in 10.75 mL of 0.02 M NaOH, diluted to 250 mL with distilled water.

▲ Universal indicator.

▲ Aluminum-copper wire, Al–Cu: If the Al wire is smooth, clean it with steel wool, dip it in 0.1 M HCl, and rinse it with water. If it is rough, dip it in acid and then rinse it in water. Twist the Cu wire around the Al wire as shown:

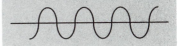

▲ For each of the indicators used, the presence of Al^{3+} ions is indicated by the red spots at the ends of the Al wire. Aluminon develops a red color in the presence of aluminum ions. The structural formula of aluminon is:

▲ In the vicinity of the Cu wire, the indicator changes to its basic color because OH^- ions are formed in this region. The indicators, their pH ranges, and their colors are as follows:

Indicator	pH Range	Color
Turmeric	7.4–8.6	yellow to red
Universal	7.0–11.0	green to purple
Thymol blue	8.0–9.6	yellow to blue
Metacresol purple	7.4–9.0	yellow to purple

Reactions

$$4Al(s) + 3O_2(g) + 6H_2O(l) \rightarrow 4Al^{3+}(aq) + 12OH^-(aq)$$

Notes

1. It takes about 2 h to prepare the agar, mix the solutions, and pour the plates. You cannot let the agar sit until some other time, because it will harden.

2. During the pouring of the plates, keep the beaker containing the agar solution in the beaker of hot water on the hot plate–stirrer. That way you do not have to be concerned with having to work fast.

3. A volume of 200 mL of agar solution is sufficient for four plates with a little to spare. It is better to have too much, rather than not enough. You need to be able to nicely cover the bottom of the plate and then cover the metal afterwards.

4. Because OH⁻ ions are produced, turmeric undergoes a pH change in the range 7.4–8.6, and the color change is from yellow to red. This will not allow you to distinguish the anodic from cathodic areas, but it is still interesting to include among the plates. Ask your students to explain "what the problem is" with this plate. (Aluminon in the presence of Al^{3+} gives a red color as well.)

5. Obvious color changes will require longer to develop if the Cu wire is not bent around the Al wire—for example, if a straight piece of Al wire is used. Try some of the variations that Orr (1949) described.

References

Masterton, W. L.; Slowinski, E. J.; Stanitski, C. L. *Chemical Principles*, alternate ed.; CBS College: Philadelphia, PA, 1983; pp 480–482.

Nebergall, W. H.; Schmidt, F. C.; Holtzclaw, H. F., Jr. *College Chemistry*, 5th ed.; Heath: Lexington, MA, 1976; pp 904–905.

Orr, W. L. *J. Chem. Educ.* **1949**, *26*, 267–268.

Concentration and Current Measurement

A series of dilute solutions of hydrochloric acid is prepared. The current of the various solutions is recorded. A direct relationship is obtained between the concentration of the solution and its current.

Procedure

Please consult the Safety Information before proceeding.

1. Using a multimeter, place a red patch cord in "A" and a black one in "COM". Attach the red patch cord to the red wire of the light-emitting diode (LED) apparatus using an alligator clip, and attach the black patch cord to the black wire of the LED apparatus.

2. Set the multimeter on 200 mA. If necessary, press the DC/AC button so the screen shows "AC".

3. Place 40 mL of solution in a 50-mL beaker.

Materials

▲ Conductance device: See annex to this investigation.

▲ Hydrochloric acid, 1.00 M HCl: Adding acid to water, mix 84.3 mL of concentrated HCl with enough water to make 1000 mL of solution.

▲ Dilute solutions of hydrochloric acid: Using a graduated pipet, add and mix the specified volume of 1.00 M HCl with enough water to make 100 mL of solution. (See table on page 288.)

▲ Multimeter.

4. Plug in the adapter.

5. Insert the probes of the LED apparatus in the solution for 10 s to a specific height (see Notes). Record the current.

6. Rinse the probes with distilled water and wipe them dry. Determine the current two more times, washing and drying the electrodes between trials. Average the results of the three trials.

7. Repeat Steps 3–6 using the other 23 solutions.

8. Plot a graph of concentration versus current (Figure 1). Use this graph in other demonstrations (see Notes).

Concepts

▲ Hydrochloric acid is a strong acid that ionizes essentially 100% in solution to produce hydronium H_3O^+ and chloride ions.

▲ When ions are present in solution, a current will be conducted. Substances that conduct a current are called *electrolytes*. The greater the concentration of ions, the greater the amount of current that will be conducted and the greater the amount of conductance.

▲ A direct relationship exists between conductance and current. Conductance equals current divided by voltage. The current is measured with the use of a multimeter, and the voltage across the probes of the LED apparatus can also be measured by using the multimeter. From this measurement, conductance can be determined.

▲ When the conductance device described is used, the values for current range from 5.73 mA for 0.00100 M HCl to 34.7 mA for 1.00 M HCl.

▲ Plotting concentration versus current results in a smooth curve. Although the ionization of hydrochloric acid is supposed to be 100%, the curve does not illustrate this principle over most of the range studied but only up to a concentration of approximately 0.011 M. Calculation of the ionization constant K_a at this concentration results in

$$K_a = \frac{(0.011 \text{ M})(0.011 \text{ M})}{(1.00 \text{ M} - 0.011 \text{ M})} = 1.22 \times 10^{-4}$$

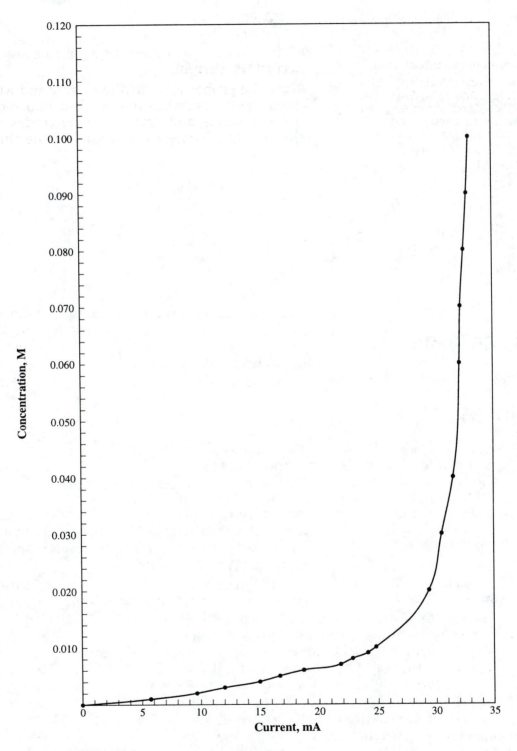

Figure 1.

▲   It appears from plotting concentration versus
current that, below a pH of approximately 1.96,
HCl is not 100% ionized (Figure 2).

Concentration (M)	Volume of 1.00 M HCl (mL)	Concentration (M)	Volume of 1.00 M HCl (mL)
0.150	15.0	0.0300	3.00
0.140	14.0	0.0200	2.00
0.130	13.0	0.0100	1.00
0.120	12.0	0.0090	0.90
0.110	11.0	0.0080	0.80
0.100	10.0	0.0070	0.70
0.090	9.0	0.0060	0.60
0.080	8.0	0.0050	0.50
0.070	7.0	0.0040	0.40
0.060	6.0	0.0030	0.30
0.050	5.0	0.0020	0.20
0.040	4.0	0.0010	0.10

Reactions

$$HCl(aq) + H_2O(l) \rightarrow H_3O^+(aq) + Cl^-(aq)$$

Notes

1. Although making the device for this demonstration involves some electrical knowledge, once
   the adapter–LED device is made, it can be used
   for a number of other investigations, such as
   34, 35, 58, and 78. If necessary, enlist the help
   of a physics teacher, electrician, or some other
   person.

2. The resistor, adapter, and bicolor LED can be
   purchased from an electronic appliance store
   such as Radio Shack.

3. The choice of 40 mL of solution and a 50-mL
   beaker is somewhat arbitrary. What is necessary, though, is that the same volume of solution in the same size beaker be used each time
   and the probes submerged the same distance.

4. Determine the height to submerge the Cu probes
   so that the 1.00 M HCl will give a current reading of 34.7 mA. Submerge the probes the same
   distance each time.

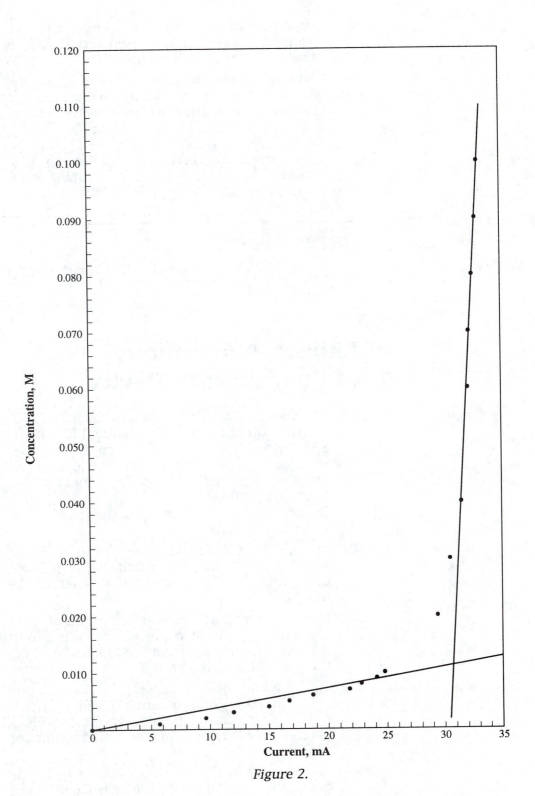

Figure 2.

5. The acid solutions can be used over and over as long as the probes are rinsed and dried after each use. Save the solutions in labeled bottles.

6. Although this demonstration was attempted with a 9-V battery and LED apparatus, the electrolysis of the water interfered in the determination of the current. Batteries are a source of dc, or direct current. It was necessary to have ac, or alternating current, to minimize the electrolysis. The ac adapter is used to obtain ac and reduce the 120 V to approximately 13 V.

References

Halliday, D.; Resnick, R. *Physics, Part II;* Wiley: New York, 1966; pp 770–813.

Annex: Preparation of Conductance Device

1. Obtain a 13-V ac adapter (Radio Shack No. 273–1610), 360-ohm resistor, bicolor LED (Radio Shack No. 276–012), about 30 cm of 14-gauge solid copper wire encased in plastic, and two alligator clips, preferably one black and one red. You will also need a thin piece of wood or plastic, about 1 cm by 13 cm, to support the wires, resistor, and LED.

2. Cut the plug off the end of the adapter and split about 3 cm of the wire in half. Remove about 0.5 cm of the plastic casing from each of the pieces.

3. Make two small holes in the top of the piece of wood that the two pieces of adapter wire can fit through. About 3 cm from the top, make a hole that the LED can fit through.

4. Put both ends of the adapter wire through one hole. Then bring one piece up through the other hole. This piece will be attached to the black patch cord of the multimeter with an alligator clip.

5. Put the LED in the hole about 3 cm from the top of the piece of wood. Solder the end of the other adapter wire to one part of the LED.

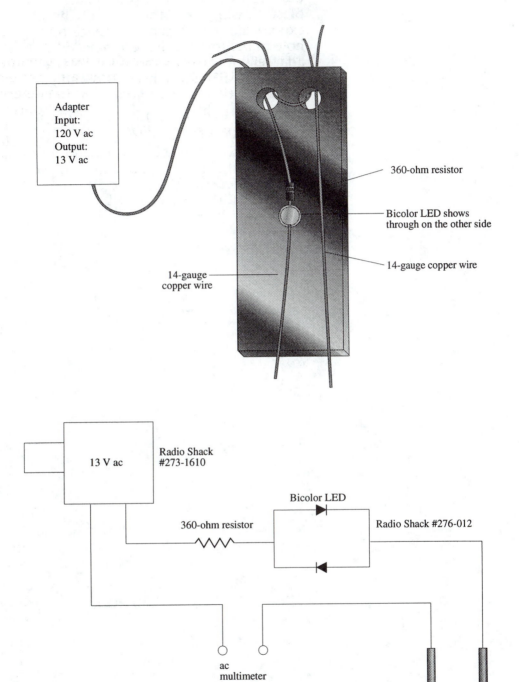

Conductance device sketch (top) and schematic (bottom). The LED apparatus was developed by Steve Spadafore of Franklin and Marshall College.

6.   Cut the 14-gauge Cu wire in half. Remove the plastic casing from the ends of the pieces to expose the wire. Insert one piece through the hole in the top of the piece of wood where the adapter wire is, with about 1 cm left at the top. This wire will be attached to the red patch cord of the multimeter with an alligator clip.

7.   Solder one end of the other piece of Cu wire to the other part of the LED.

8.   At the end of the piece of wood, opposite to where the holes are, glue the two pieces of Cu wire in place with silicone so approximately 1–1.5 cm of Cu wire is exposed.

Conductance of Various Substances

Measuring the current, in milliamperes, of various substances, both molecular and ionic, will give an indication of the presence or lack of ions.

Procedure 1: For All Solutions

Please consult the Safety Information before proceeding.

1. Using a multimeter, place a red patch cord in "A" and a black one in "COM". Attach the red patch cord to the red wire of the light-emitting diode (LED) apparatus (see Investigation 77) using an alligator clip, and attach the black patch cord to the black wire of the LED apparatus.

2. Set the multimeter on 200 mA. If necessary, press the DC/AC button so the screen shows "AC".

3. Acetic acid is harmful if swallowed, inhaled, or absorbed through the skin. It is destructive to the tissue of the mucous membranes and the upper respiratory tract, eyes, and skin.

4. Silver nitrate is caustic and irritating to the skin and mucous membranes and can cause burns. It is highly toxic by ingestion.

3. Place 40 mL of solution in a 50-mL beaker.

4. Plug in the adapter.

5. Insert the probes of the LED apparatus in the solution for 10 s to a specific height (see Notes). Record the current.

6. Rinse the probes with distilled water and wipe them dry. Determine the current two more times, washing and drying the electrodes between trials. Average the results of the three trials.

7. Repeat Steps 3–6 using the other seven solutions.

Procedure 2: For Molten Candle Wax and Silver Nitrate

1. Place a crucible containing solid candle wax on a hot plate at low heat.

2. Melt the solid.

3. Gently heat the electrodes on the edge of the hot plate for a few seconds.

4. Place the electrodes in the molten substance and record the value.

5. Remove the electrodes, disconnect the power source, and clean the electrodes by scraping the solid off with a spatula. Rinse the electrodes with distilled water and wipe them dry.

6. Remove the crucible from the hot plate so the molten substance can solidify.

7. Repeat Steps 1–6 with a crucible of silver nitrate instead of wax. Exercise caution with the silver nitrate. **DO NOT** allow any student to do this part of the demonstration. The electrodes must be dry to prevent any molten compound from being emitted.

Concepts

▲ For a substance to conduct a current, ions must be present. Substances that dissolve in water to form solutions that conduct a current are called *electrolytes*. A solution of an ionic compound, such as sodium chloride, sodium hydroxide, or silver nitrate, will conduct a current.

▲ Most molecular substances, whether molten or in solution, do not conduct a current. These substances are called *nonelectrolytes*. Examples include candle wax, glucose, starch, and alcohols.

▲ Molecular acids are an exception. They conduct when in solution, because of the presence of ions. Examples include hydrochloric and acetic acid.

▲ A solid ionic substance will not conduct because its ions are nonmobile. When molten, or melted, an ionic compound will conduct. An example is silver nitrate.

▲ Following are the results based on the device used (solutions are all 1.0 M):

Substance	Current (mA)
Starch (aqueous)	1.0
Glucose (aqueous)	0.8
1-Propanol (liquid)	0.0
Distilled water (liquid)	1.1
HCl (aqueous)	34.7
CH₃COOH (aqueous)	17.6
NaOH (aqueous)	34.5
NaCl (aqueous)	33.6
AgNO₃ (aqueous)	34.5
Candle wax (liquid)	0.0
AgNO₃ (liquid)	difficult to determine precisely

▲ The concentration of an ionic solution will determine how much current is conducted. The higher the concentration, the more ions and the greater the amount of current conducted.

Materials

▲ For the following 1.0 M solutions, dissolve or mix the specified amount in enough water to make 100 mL of solution:

1. Glucose, $C_6H_{12}O_6$: 18.02 g.
2. Silver nitrate, $AgNO_3$: 16.99 g.
3. Sodium hydroxide, NaOH: 4.00 g.
4. Sodium chloride, NaCl: 5.84 g.
5. Concentrated acetic acid, $HC_2H_3O_2$: 5.8 mL.
6. Concentrated hydrochloric acid, HCl: 8.4 mL.
7. Concentrated aqueous ammonia, NH_3: 6.8 mL.

▲ Small crucible, not a beaker, of solid candle wax.

▲ Small crucible of silver nitrate.

▲ Multimeter.

▲ LED apparatus.

Reactions

$$HCl(aq) + H_2O(l) \rightarrow H_3O^+(aq) + Cl^-(aq)$$
$$HC_2H_3O_2(aq) + H_2O(l) \leftrightarrows H_3O^+(aq) + C_2H_3O_2^-(aq)$$
$$NaOH(aq) \rightarrow Na^+(aq) + OH^-(aq)$$
$$NaCl(aq) \rightarrow Na^+(aq) + Cl^-(aq)$$
$$AgNO_3(aq) \rightarrow Ag^+(aq) + NO_3^-(aq)$$

$$AgNO_3(s) \overset{\Delta}{\leftrightarrows} Ag^+(l) + NO_3^-(l)$$

Notes

1. Other acids, alcohols, or ionic compounds in solution could be used.

2. The choice of 40 mL of solution and a 50-mL beaker is somewhat arbitrary. What is necessary, though, is that the same volume of solution in the same size beaker be used each time and the probes submerged the same distance (consult Notes in Investigation 77).

3. Because the LED apparatus is used in several demonstrations, its use is also suggested in this demonstration, but a lightbulb conductivity apparatus using a 150-W unfrosted lightbulb could be used and the degree of brightness monitored.

Gallium Beating Heart

Liquid gallium is placed in a Petri dish containing sulfuric acid and dichromate ions. When it is touched with an iron nail, the gallium begins to pulsate.

Procedure

Please consult the Safety Information before proceeding.

1. Place about 15 g of gallium in a glass Petri dish.

2. Add about 50 mL of 6 M sulfuric acid to the dish and very carefully warm the mixture to 55 °C on a hot plate.

3. After the sulfuric acid solution has reached 55 °C and the gallium has assumed a dull appearance and flat shape, place the dish on an overhead projector.

3. Potassium dichromate is a strong oxidizing agent that is harmful to the skin, mucous membranes, and eyes. Avoid inhalation of the vapors.

4. Add 1 mL of the potassium dichromate solution to the Petri dish.

5. Place a nail directly over the center of the blob of gallium; gently touch it to start it beating.

6. Adjust the position of the nail to produce continuous beating.

7. Clamp or hold the nail in this position.

Concepts

▲   The gallium serves as an electron switch for the oxidation–reduction reaction.

▲   The iron of the nail is oxidized by the acid. The surface of the nail acquires an excess of electrons that will be transferred to the gallium when they touch.

▲   The dichromate oxidizes the surface atoms of the gallium and forms a coating of gallium sulfate.

▲   The formation of the gallium sulfate apparently interferes with the surface tension. The surface tension decreases, and the gallium loses its spherical shape and flattens out.

▲   When the gallium touches the iron nail, the electrons are transferred to the gallium sulfate. The gallium(III) ion gains three electrons and forms gallium metal, restoring the spherical shape. The process repeats for about a half hour.

▲   The potentials for the four major reactions give a final positive value. The $Fe^0|\ Ga^{3+}|\ |\ Fe^{3+}|\ Ga^0$ reaction is negative and the $Ga^0|\ Cr^{6+}|\ |\ Ga^{3+}|\ Cr^{3+}$ is positive, resulting in a positive final voltage. See Ealy (1993) for a more complete discussion.

$$Fe^0 \rightarrow Fe^{2+} + 2e^- \quad +0.441 \text{ V}$$
$$Fe^{2+} \rightarrow Fe^{3+} + e^- \quad -0.770 \text{ V}$$
$$Ga^0 \rightarrow Ga^{3+} + 3e^- \quad +0.520 \text{ V}$$
$$3e^- + Cr^{6+} \rightarrow Cr^{3+} \quad\quad +1.10 \text{ V}$$

Reactions

1. $Fe(s) + 2H^+(aq) \rightarrow Fe^{2+}(aq) + H_2(g)$

2. $14H^+(aq) + 6Fe^{2+}(aq) + Cr_2O_7^{2-}(aq) \rightarrow$
$$6Fe^{3+}(aq) + 2Cr^{3+}(aq) + 7H_2O(l)$$

3. $2Ga^0(aq) \rightarrow 2Ga^{3+}(aq) + 6e^-$

4. $6e^- + 14H^+(aq) + Cr_2O_7^{2-}(aq) \rightarrow 2Cr^{3+}(aq) + 7H_2O(l)$

5. $2Ga(s) + 14H^+(aq) + Cr_2O_7^{2-}(aq) \rightarrow$
$$2Cr^{3+}(aq) + 2Ga^{3+}(aq) + 7H_2O(l)$$

6. $Ga^{3+}(aq) + Fe(s) \rightarrow Fe^{3+}(aq) + Ga(s)$

7.(a) $6e^- + 2Ga^{3+} \rightarrow 2Ga^0$ -0.520

  (b) $2Fe^0 \rightarrow 2Fe^{2+} + 4e^-$ $+0.441$

  (c) $2Fe^{2+} \rightarrow 2Fe^{3+} + 2e^-$ -0.770

  (d) $2Ga^{3+} + 2Fe^0 \rightarrow 2Fe^{3+} + 2Ga^0$ -0.849 **-0.849**

8.(a) $2Ga^0 \rightarrow 2Ga^{3+} + 6e^-$ $+0.520$

  (b) $2Cr^{6+} + 6e^- \rightarrow 2Cr^{3+}$ $+1.10$

  (c) $2Cr^{6+} + 2Ga^0 \rightarrow 2Cr^{3+} + 2Ga^{3+}$ $+1.62(+)$ **$+1.62$**

      cell emf: **$+0.77$**

Materials

▲   Sulfuric acid, 6 M H_2SO_4: Carefully add 36 mL of concentrated acid to 64 mL of distilled water and stir the solution.

▲   Potassium dichromate, 0.1 M $K_2Cr_2O_7$: Dissolve 3 g of $K_2Cr_2O_7$ in enough water to make 10 mL of solution. Store it in a tightly closed bottle until needed again. When presenting this demonstration, add 1 mL of this solution to 50 mL of 6 M sulfuric acid in the Petri dish. This dilution will make the dichromate ion about 0.002 M.

▲   Gallium metal, Ga.

Notes

1. When the nail touches the surface of the gallium, it should jerk away and form a spherical shape with a very shiny surface. If the nail is held against the gallium too long, the gallium will become very viscous. You will have to wait until it pops back to a spherical shape to begin the beating.

2. As the gallium cools to room temperature, the beating will slow. However, the gallium does not freeze. The heat produced by the oxidation and reduction will keep the gallium melted.

References

Adams, R. H. *J. Chem. Educ.* **1933**, *10*, 512.

Avir, D. *J. Chem. Educ.* **1989**, *66*, 211.

Campbell, J. A. *J. Chem. Educ.* **1957**, *34*, 363.

CRC Handbook of Chemistry and Physics, 48th ed.; Weast, R. C., Ed.; CRC: Cleveland, OH, 1967; p D86.

Ealy, J. L. *J. Chem. Educ.* **1993**, *70*, 491.

Keizer, J.; Rock, P. A.; Lin, S.-W. *J. Am. Chem. Soc.* **1979**, *101*, 5637.

Lin, S.-W.; Keizer, J.; Rock, P. A.; Stenschke, H. *Proc. Natl. Acad. Sci. U.S.A.* **1974**, *71*, 4471.

Current Flow Direction

When a bicolor diode is placed across the terminals of an electrochemical cell, the direction of the electron flow can be easily established by the green or red glow of the diode.

Procedure

Please consult the Safety Information before proceeding.

1. Set up three electrochemical cells in series as follows. Place six large (25×150 mm) test tubes in a test-tube rack. Pour 1.0 M $CuSO_4$ into the first, third, and fifth test tubes and 1.0 M $ZnSO_4$ into the second, fourth, and sixth test tubes to about the three-fourths mark. Place copper and zinc electrodes into their respective solutions. Place a 10-cm piece of KCl-soaked paper towel

Safety Information

1. Copper(II) sulfate pentahydrate may be harmful by inhalation, ingestion, or skin absorption and may cause eye and skin irritation. It may be irritating to mucous membranes and the upper respiratory tract.

2. Zinc sulfate heptahydrate and potassium chloride are considered to be nonhazardous; however, all of their characteristics may not have been investigated yet.

Materials

▲ Copper(II) sulfate penta-hydrate, 1.0 M $CuSO_4 \cdot 5H_2O$: Add 250 g to enough water to make 1.0 L of solution.

▲ Zinc sulfate heptahy-drate, 1.0 M $ZnSO_4 \cdot 7H_2O$: Add 287 g to enough water to make 1.0 L of solution.

▲ Potassium chloride, satu-rated KCl: Add 10 g to 25 mL of the water and stir. Place three 10-cm pieces of rolled-up paper towel in this solu-tion.

▲ Multimeter.

▲ Bicolor diode.

▲ 9-V battery.

from the first to the second tube. Repeat with the third and fourth and with the fifth and sixth tubes. The towel must extend down into the solution. Heavy butcher cord can be substituted for the paper towels. Connect the zinc electrode in the second test tube to the copper electrode with a patch cord and repeat with the remaining cells. Connect a patch cord to the copper electrode in the first test tube and connect another patch cord to the zinc electrode in the sixth cell. These two patch cords will be connected to the bicolor diode.

2. Set the multimeter on the 0–15-V scale. Check the voltage of the three cells with a multimeter. When the red (+) lead is touched to the copper electrode in the first tube and the black (–) lead is touched to the zinc electrode in the sixth tube simultaneously, the multimeter should read about 3.3 V. When the leads are reversed, the meter should show a negative reading or a reading below 0.0 V.

3. Touch the multimeter leads to a 9-V battery to show that the meter reads a positive value when the red (+) lead is touched to the positive terminal and the black (–) lead is touched to the negative terminal.

4. Notice that one of the diodes leads is longer than the other. Connect two patch cords to the bicolor diode and touch the free clips to the 9-V battery. Record which color diode glows when the longer lead is connected to the positive side of the battery and which color diode glows when the longer lead is connected to the negative side of the battery. The diode should glow green when the longer lead is connected to the positive side of the battery. The diode should glow red when the longer lead is connected to the negative side of the terminal.

5. When you connect the longer lead of the diode to the patch cord attached to the copper electrode of the first test tube and the shorter lead to the patch cord attached to the zinc electrode of the sixth tube, the diode should glow green. The room may need to be darkened to observe the faint glow. Reverse the diode leads and the diode should now glow green.

Concepts

▲ In electrochemical cells, oxidation and reduction take place at or on the surface of the electrodes. The standard cell potentials indicate which elements will be oxidized or reduced. With a Zn | Cu cell, copper's potential is 0.34 V and zinc's potential is –0.76 V. The zinc has a negative reduction potential and will be oxidized more easily than the copper with a positive reduction potential.

▲ The cell produces a potential of 1.1 V, with the electrons flowing from the site of oxidation (Zn electrode) to the site of reduction (Cu electrode).

▲ *Standard* current flows in the opposite direction of *electron* current. Standard current can be loosely described as the apparent movement of a "positive hole" in a direction opposite to the apparent flow of the negative electrons. These positive holes are the places into which the moving negative electrons go. Many older physics books use only standard current, and therefore many unnecessary discussions and conflicting opinions have ensued over the direction of current flow in cells and batteries. These correct but differing statements have resulted in unnecessary confusion for students.

▲ Electrons are lost from the zinc metal as it is oxidized to zinc ions. These electrons are left on the zinc electrode. Zinc metal is a good conductor, and the electrons can travel through the electrode and through the bicolor diode on to the copper electrode. At the copper electrode, the electrons combine with the copper(II) ions, which reduces them to copper atoms.

▲ The KCl-soaked towel acts as a salt bridge to maintain electrical neutrality in each half-cell. In the half-cell where oxidation takes place, additional Zn^{2+} ions are generated in the $ZnSO_4$ solution. Negative ions, Cl^-, flow from the salt bridge into the copper half-cell to maintain a charge balance. Cu^{2+} ions are consumed from the copper sulfate solution, so K^+ ions flow from the salt bridge, also to maintain a charge balance.

Reactions

$$Cu^{2+}(aq) + 2e^- \rightarrow Cu^0(s) \qquad\qquad +0.34 \text{ V}$$
$$Zn^0(s) \rightarrow Zn^{2+}(aq) + 2e^- \qquad +0.76 \text{ V}$$
$$\overline{Cu^{2+}(aq) + Zn^0(s) \rightarrow Cu^0(s) + Zn^{2+}(aq) \quad +1.10 \text{ V}}$$

References

CRC Handbook of Chemistry and Physics, 66th ed.; Weast, R. C., Ed.; CRC: Boca Raton, FL, 1985.

Synthesis

Synthesis of Mohr's Salt

Cold solutions of iron(II) sulfate and ammonium sulfate are mixed in a test tube. When the inside of the test tube is scratched, pale green crystals of Mohr's salt immediately begin to form.

Procedure

Please consult the Safety Information before proceeding.

1. Chill 10 mL of iron(II) sulfate solution and 10 mL of ammonium sulfate solution in separate test tubes in an ice bath for 5 min.

2. Pour the iron(II) sulfate solution into the ammonium sulfate solution and mix by swirling.

3. Scratch the bottom of the test tube with a stirring rod. Almost immediately, crystals begin to form.

Materials

▲   Ammonium sulfate, saturated $(NH_4)_2SO_4$: Dissolve 76 g in 100 mL of distilled water.

▲   Iron(II) sulfate heptahydrate, saturated $FeSO_4 \cdot 7H_2O$: Dissolve 70 g in 100 mL of distilled water.

4.   Collect the crystals by gravity filtration. Wash them once with ice-cold water. After drying the crystals, keep them in a well-closed bottle, protected from the light.

Concepts

▲   The double salt formed, iron(II) ammonium sulfate hexahydrate, is known as Mohr's salt. It is used in laboratory work as a reducing agent.

▲   Mohr's salt was introduced into volumetric analysis by K. F. Mohr in the 1850s. Mohr also invented the specific gravity balance, the buret, the pinch clamp, and the cork borer.

▲   Although most iron(II) salts are easily oxidized to iron(III) in the air, Mohr's salt, which is a double sulfate, is less readily oxidized. Other factors affecting the oxidation of iron(II) salts are the ligands attached and the pH in aqueous solution, alkaline solutions being very readily oxidized.

▲   This demonstration can be used to illustrate the ease with which a compound can be synthesized, and a comparison can be made between the oxidation in air of iron(II) sulfate hexahydrate and the newly synthesized iron(II) ammonium sulfate hexahydrate.

Reactions

$$2NH_4^+(aq) + 2SO_4^{2-}(aq) + Fe^{2+}(aq) + 6H_2O(l) \rightarrow$$
$$(NH_4)_2SO_4 \cdot FeSO_4 \cdot 6H_2O(s)$$

Notes

1.   The solubility of the original two solids can be compared with that of the iron(II) ammonium sulfate hexahydrate to show the difference among the solids. At 15 °C the solubilities, in grams per 100 mL of water, are as follows:

$FeSO_4 \cdot 7H_2O$	70
$(NH_4)_2SO_4$	74.5
$(NH_4)_2SO_4 \cdot FeSO_4 \cdot 6H_2O$	29

2. Melting points could also be compared. A hot oil bath could be used for the ammonium sulfate. The values are as follows: iron(II) sulfate heptahydrate, 90 °C; ammonium sulfate, 235 °C; and iron(II) ammonium sulfate hexahydrate, 100–110 °C.

References

Fowles, G. *Lecture Experiments in Chemistry;* Blakiston: Philadelphia, PA, 1937; pp 231–232.

Greenwood, N. N.; Earnshaw, A. *Chemistry of the Elements,* 1st ed.; Pergamon: New York, 1984; p 1269.

Nebergall, W. H.; Schmidt, F. C.; Holtzclaw, H. F., Jr. *College Chemistry,* 5th ed.; Heath: Lexington, MA, 1976; pp 907–908.

White, Needlelike Crystals of Boric Acid

Dropwise addition of hydrochloric acid to a solution of sodium borate results in the formation of white needlelike crystals.

Procedure

Please consult the Safety Information before proceeding.

1. Heat 3 mL of water in a test tube in a boiling water bath.

2. Gradually add 4.5 g of sodium borate decahydrate, stirring to dissolve.

3. Dropwise, add up to 20 drops of concentrated hydrochloric acid, noting the needlelike crystals that form on the surface and gradually fall to the bottom.

4. Cool the test tube in an ice bath for further crystal formation.

Safety Information

1. Even though sodium borate decahydrate is used as an ingredient for washing mucous membranes, ingestion of 5–10 g by young children can cause severe vomiting, diarrhea, shock, and death.

2. External contact with concentrated hydrochloric acid causes severe burns, and contact with the eyes may result in a total loss of vision. Inhalation causes coughing and choking with possible inflammation and ulceration of the respiratory tract. Work in a fume hood and wear goggles, gloves, and a rubber apron.

Materials

▲ Sodium borate deca-hydrate, $Na_2B_4O_7 \cdot 10H_2O$, powder.

▲ Hydrochloric acid, concentrated HCl: commercial stock solution.

Concepts

▲ The molecular shape of boric acid, H_3BO_3, is an equilateral triangle with the boron atom at the center and the oxygen atoms at the corners. The triangular units are held together by hydrogen bonding in the solid acid.

▲ One of the oldest and most complex processes of the chemical industry is crystallization. Factors such as heat and mass transfer, fluid and particle mechanics, a thermodynamically unstable solution in which the solids are suspended, and traces of impurities can all affect nucleation and crystal growth.

▲ Different crystal faces grow at different rates under different environmental conditions. Changes in the face-growth rates give rise to habit (shape) changes in crystals. Changes in the environment such as temperature, supersaturation, pH, impurities, and the solvent can profoundly affect individual face-growth rates. For instance, sodium borate decahydrate's normal habit is needles, but in the presence of (carboxymethyl)cellulose, flakes form.

▲ Careful control of crystal size and habit is important in the commercial production of soda ash, Na_2CO_3, which is prepared from natural deposits of the mineral trona, $Na_2CO_3 \cdot NaHCO_3 \cdot 2H_2O$, found in the Green River basin of Wyoming. The annual production of soda ash in the United States exceeds 17 billion pounds, and it is used in the manufacture of glass, pulp, soaps, textiles, and detergents.

▲ Heating boric acid to 100 °C forms metaboric acid, HBO_2. Further heating to 140–160 °C forms tetraboric acid, $H_2B_4O_7$. At higher temperatures, boric oxide, B_2O_3, is formed.

▲ Borax is made from boric acid by adding anhydrous sodium carbonate to a boiling solution of the acid. Carbon dioxide gas is evolved, and the borax is crystallized with water to form sodium borate decahydrate.

▲ In a purified form, boric acid is an antiseptic and astringent. It is also used in glassmaking and as a preservative agent for perishable articles.

Reactions

$$B_4O_7{}^{2-}(aq) + 5H_2O(l) + 2H^+(aq) \rightarrow 4H_3BO_3(s)$$

Notes

1. The crystals can be collected by gravity filtration.

2. Test the solubility of the crystals in water, ethanol, and HCl.

References

Kingzett, C. T. *Chemical Encyclopedia;* Van Nostrand: New York, 1928; p 83.

McQuarrie, D. A.; Rock, P. A. *General Chemistry*, 2nd ed.; Freeman: New York, 1987; p 85.

Mullin, J. W. In *Kirk–Othmer Encyclopedia of Chemical Technology*, 3rd ed.; Wiley: New York, 1979; Vol. 7, pp 243–253.

Nebergall, W. H.; Schmidt, F. C.; Holtzclaw, H. F., Jr. *College Chemistry*, 5th ed.; Heath: Lexington, MA, 1976; pp 725–726.

Newth, G. S. *Chemical Lecture Experiments;* Longmans, Green, & Co.: New York, 1892; p 226.

Formation of Black Copper(II) Oxide

Warming a blue precipitate of copper(II) hydroxide causes it to turn green, greenish blue, and then black, forming copper(II) oxide.

Procedure

Please consult the Safety Information before proceeding.

1. Place 25 mL of potassium hydroxide solution in a 250-mL beaker, and put the beaker on a hot plate–stirrer.

2. Add 22 mL of copper(II) sulfate solution.

3. Turn the hot plate on low temperature and the stirrer fast enough to stir the greenish-blue gelatinous precipitate, which is copper(II) hydroxide. Heat until the precipitate turns black. Observe the color changes.

Safety Information

1. Copper(II) sulfate pentahydrate is toxic if ingested and may be harmful if it comes in contact with mucous membranes or if the dust is inhaled.

2. Potassium hydroxide is harmful if swallowed, inhaled, or absorbed through the skin. The material is extremely destructive to the tissue of the mucous membranes and the upper respiratory tract, eyes, and skin. Wear gloves when handling.

Materials

▲   Copper(II) sulfate penta-hydrate, 0.5 M $CuSO_4 \cdot 5H_2O$: Dissolve 1.5 g in enough water to make 100 mL of solution.

▲   Potassium hydroxide, 1.0 M KOH: Dissolve 5.6 g in enough water to make 100 mL of solution.

4.   Repeat Steps 1–3, except in Step 1 bring the potassium hydroxide to a boil and then add the copper sulfate solution. Observe the immediate formation of the black precipitate.

Concepts

▲   The addition of hydroxide ions to cold solutions of Cu^{2+} produces a greenish-blue gelatinous precipitate of copper(II) hydroxide.

▲   Heating a solution containing copper(II) hydroxide will result in the formation of black copper(II) oxide.

▲   The gelatinous precipitate of copper(II) hydroxide is a colloidal suspension known as a sol. In a sol, the dispersion medium is a liquid and the dispersed phase is a solid. Other examples of sols are paints, some inks, starch suspension, and milk of magnesia. Using a He–Ne laser, the presence of a sol can be demonstrated by observing the laser beam as it is scattered by the gelatinous precipitate. Exercise extreme caution if a laser is used. Do not point it at anyone's eyes.

▲   Other metal hydroxides, besides copper, are also thermally unstable. Heating cobalt(II) hydroxide in the absence of air produces olive-green CoO, and oxidation in air with dehydration of manganese(II) hydroxide produces black Mn_2O_3.

▲   Iron(II) hydroxide gradually decomposes to $FeO \cdot Fe_2O_3$ with release of hydrogen. This hydroxide, in the presence of oxygen, darkens rapidly and eventually forms reddish-brown hydrated iron(III) oxide. Iron(III) hydroxide produces a gelatinous reddish-brown precipitate of hydrated oxide, which forms Fe_2O_3 when heated to 200 °C.

▲   Adding alkali to aqueous solutions of Ni(II) salts and heating produces green NiO, and soluble silver(I) salts treated with alkali produce dark brown precipitates of Ag_2O.

Reactions

1. $Cu^{2+}(aq) + 2OH^-(aq) \rightarrow Cu(OH)_2(s)$
                                    blue–green

2. $Cu(OH)_2(s) \overset{\Delta}{\to} CuO(s) + H_2O(l)$
        black

Notes

1.   You can also use 1.0 M NaOH in place of potassium hydroxide.

2.   If you initially filter some of the greenish-blue gelatinous precipitate, it will not stay blue very long after exposure to the air. You can, however, filter the black precipitate at the end, and it will stay black.

References

Fowles, G. *Lecture Experiments in Chemistry;* Blakiston: Philadelphia, PA, 1937; pp 448–449.

Greenwood, N. N.; Earnshaw, A. *Chemistry of the Elements*, 1st ed.; Pergamon: New York, 1984; pp 1218, 1254–1255, 1296, 1336, 1373.

Nebergall, W. H.; Schmidt, F. C.; Holtzclaw, H. F., Jr. *College Chemistry*, 5th ed.; Heath: Lexington, MA, 1976; pp 318–324, 870.

Potassium Tris(oxalato)chromate(III) Trihydrate

Potassium dichromate solid is added to a hot solution of oxalic acid. After the solution comes to a boil, potassium oxalate monohydrate is added, the solution is cooled, and upon addition of ethanol, dark blue–green crystals of potassium tris(oxalato)chromate(III) trihydrate form.

Procedure

Please consult the Safety Information before proceeding.

1. Using a tall-form 400-mL beaker on a hot plate–stirrer, with stirring dissolve 9 g of oxalic acid dihydrate, HOOCCOOH·2H$_2$O, in 20 mL of warm water.

2. In small portions, add 3 g of potassium dichromate, K$_2$Cr$_2$O$_7$, waiting each time until the vigorous reaction stops before adding the next portion.

Safety Information

1. Oxalic acid is caustic and corrosive to the skin and mucous membranes. Ingestion may cause severe gastroenteritis. Inhalation of the dust can be harmful.

2. Potassium oxalate monohydrate is poisonous.

3. Potassium dichromate is a strong oxidizing agent that is harmful to the skin, mucous membranes, and eyes. Avoid inhalation of the vapors.

Materials

▲ Oxalic acid dihydrate, HOOCCOOH·2H$_2$O, powder.

▲ Potassium dichromate, K$_2$Cr$_2$O$_7$, powder.

▲ Potassium oxalate monohydrate, K$_2$C$_2$O$_4$·H$_2$O, powder.

▲ Ethanol, C$_2$H$_5$OH.

3. Heat to boiling and add 3.5 g of potassium oxalate monohydrate, stirring until dissolved.

4. Remove the beaker from the heat and cool the solution for 10 min.

5. Add 4 mL of ethanol, C$_2$H$_5$OH, with stirring. Observe as dark blue-green crystals form.

Concepts

▲ Most transition metal compounds are colored. The d orbitals in the compounds in any one energy level are not of the same energy but are commonly split into sets of orbitals separated by energies corresponding to wavelengths in the visible region. Electronic transitions occur between orbitals in these sets when visible light is absorbed. The complementary color of light is transmitted and that is what we see. For example, absorbance of yellow light causes the compound to appear blue.

▲ Vacant d orbitals in transition metals can accommodate an electron pair that is shared and provided by some other species, and this feature results in the formation of a coordinate covalent bond. Chromium(III) forms stable salts with many species such as CN$^-$, SCN$^-$, N$_3^-$, C$_2$O$_4^{2-}$, H$_2$O, and NH$_3$. All of these species are capable of donating an electron pair. With almost no exceptions the complexes are hexacoordinate and octahedral and may be negative, positive, or neutral.

▲ Oxalic acid is a dicarboxylic acid found as the calcium or potassium salt in wood sorrel, rhubarb, and many lichens and fungi. It imparts a sour taste to plants. The mold *Aspergillus niger* and many species of *Penicillium* metabolize sugar solutions into calcium oxalate. The structure of oxalic acid is

$$\underset{\text{HOC—COH}}{\overset{\text{O O}}{\underset{\|\ \ \|}{}}}$$

Reactions

$$3K^+(aq) + Cr^{3+}(aq) + 3C_2O_4^{2-}(aq) + 3H_2O(l) \rightarrow$$
$$K_3Cr(C_2O_4)_3 \cdot 3H_2O(s)$$

Notes

1. The crystals can be collected by vacuum filtration. Wash the crystals with an equal-volume mixture of water and ethanol, and then wash with ethanol.

2. The demonstration could also be done microscale in a test tube. The test tube could be suspended in a hot water bath, using a stirring rod to stir.

References

Greenwood, N. N.; Earnshaw, A. *Chemistry of the Elements;* Pergamon: New York, 1984; pp 1196–1200.

Kingzett, C. T. *Chemical Encyclopedia;* Van Nostrand: New York, 1928; pp 521–522.

Nebergall, W. H.; Schmidt, F. C.; Holtzclaw, H. F., Jr. *College Chemistry;* Heath: Lexington, MA, 1976; p 673.

Palmer, W. G. *Experimental Inorganic Chemistry;* University Press: Cambridge, MA, 1954; pp 386–387.

Whitten, K. W.; Gailey, K. D. *General Chemistry with Qualitative Analysis*, 2nd ed.; Saunders: Philadelphia, PA, 1984; pp 781–782.

Iron(II) Oxalate Dihydrate

Iron(II) ammonium sulfate hexahydrate is dissolved in warm acidified water. After boiling, addition of oxalic acid solution results in the formation of a yellow solid, iron(II) oxalate dihydrate.

Procedure

Please consult the Safety Information before proceeding.

1. Using a hot plate–magnetic stirrer, warm 25 mL of water acidified with 2 drops of concentrated sulfuric acid in a 150-mL beaker. With stirring, add and dissolve 7.5 g of iron(II) ammonium sulfate hexahydrate.

2. Continue to heat the solution to boiling with stirring and add 37.5 mL of oxalic acid solution. A yellow solid will form during the boiling.

Safety Information

1. Concentrated sulfuric acid is very corrosive. Handle with caution and avoid contact with the skin because it produces severe burns. Inhalation of the concentrated vapor may cause serious lung damage. Work in a fume hood, and wear a rubber apron, gloves, and goggles.

2. Oxalic acid is caustic and corrosive to the skin and mucous membranes. Ingestion may cause severe gastroenteritis. Inhalation of the dust can be harmful. →

3. Iron(II) ammonium sulfate hexahydrate may be harmful by inhalation, ingestion, or skin absorption. It causes eye and skin irritation. It is irritating to mucous membranes and the upper respiratory tract.

3. Remove the mixture from the heat and allow the solid to settle. Decant the clear supernatant liquid and discard.

4. Add 25 mL of hot water. Stir the mixture. Decant the clear supernatant liquid and discard it. Observe the yellow solid.

Concepts

▲ Iron(II) ammonium sulfate hexahydrate, known as Mohr's salt, was first used for volumetric analysis by K. F. Mohr in the 1850s. Most Fe(II) salts are unstable and undergo oxidation in air to Fe(III). Double sulfates, such as this salt, are more stable than other Fe(II) salts.

▲ Fe(II) salts in acidic solution, as in the demonstration, are also much less susceptible to oxidation.

▲ Addition of oxalate ions, $C_2O_4^{2-}$, to the acidified Fe(II) solution results in the formation of yellow iron(II) oxalate dihydrate, $FeC_2O_4 \cdot 2H_2O$.

Materials

▲ Oxalic acid dihydrate, $C_2H_2O_4 \cdot 2H_2O$: Dissolve 10 g in 100 mL of warm water.

▲ Iron(II) ammonium sulfate hexahydrate, $Fe(NH_4)_2(SO_4)_2 \cdot 6H_2O$, powder.

▲ Sulfuric acid, concentrated H_2SO_4: commercial stock solution.

Reactions

$$Fe^{2+}(aq) + C_2O_4^{2-}(aq) + 2H_2O(l) \rightarrow FeC_2O_4 \cdot 2H_2O \ (s)$$

Notes

The demonstration could be performed microscale by using a test tube in a hot water bath.

References

Greenwood, N. N; Earnshaw, A. *Chemistry of the Elements;* Pergamon: New York, 1984; p 1269.

Palmer, W. G. *Experimental Inorganic Chemistry;* University Press: Cambridge, MA, 1954; p 519.

Pyrophori

Four metal oxalates are produced from salts of the metals. The ligand strength of the oxalate ion is shown. The exothermic conversion from oxides with lower oxidation numbers to oxides with higher oxidation numbers is vividly demonstrated.

Procedure

Please consult the Safety Information before proceeding.

1. Add 9 g of iron(II) sulfate and 5 g of oxalic acid to 100 mL of water in a 150-mL beaker.

2. Place the beaker on a magnetic stirrer and stir the solution for 20 min.

3. When the solution has formed a heavy yellow precipitate, stop stirring it and filter the solution.

4. Wash the precipitate with 100 mL of distilled water.

3. Cobalt(II) nitrate hexahydrate may be harmful by inhalation, ingestion, or skin absorption and may cause eye and skin irritation. It may be irritating to mucous membranes and the upper respiratory tract.

4. Nickel(II) sulfate hexahydrate is toxic by ingestion or inhalation. It may be harmful by skin absorption and may cause eye and skin irritation. It may be irritating to mucous membranes and the upper respiratory tract.

5. Manganese(II) sulfate monohydrate may be harmful by inhalation, ingestion, or skin absorption and may cause eye and skin irritation. It may be irritating to the mucous membranes or respiratory tract.

5. Dry the precipitate overnight or for about an hour in a low-temperature oven.

6. Scrape the precipitate from the filter paper and grind it in a mortar and pestle.

7. Place about a spoonful of the precipitate in a Pyrex test tube and heat it gently.

8. Water and gas will be seen escaping from the mouth of the tube as the color changes from yellow iron(II) oxalate to black iron(II) oxide.

9. Do not heat to red hot; heat only until all of the oxalate is converted to the oxide.

10. Wait until the tube and contents have cooled, turn off the lights, and pour the contents onto the floor. The metal oxide powders will rapidly oxidize on contact with atmospheric oxygen and will release heat and light, similar to a Fourth of July sparkler. (Clean up with a vacuum cleaner.)

11. Repeat Steps 1–10 with the nickel(II), cobalt(II), and manganese(II) salts to produce their oxalates.

Concepts

▲ Examples of this behavior were called *pyrophori* by Newth in his 1892 seminal work. Today, the term *pyrophoric* has a similar connotation, especially finely divided metals that burst into flame when placed in contact with air. The pyrophori reaction should not become confused with dust explosions in grain elevators and coal dust explosions in mines. Those explosions need a spark to initiate the reaction (i.e., to reach activation energy), which then results in a very rapid combustion. Pyrophori reactions do not need a spark because room temperature is high enough to exceed the energy of activation. When the metal oxalate is precipitated, it does so as a very finely divided powder. This state of the reactants positively affects the rate of reaction.

▲ The iron oxalate dihydrate when heated decomposes into anhydrous iron oxalate, releasing water, then into iron carbonate with the release of carbon monoxide. Upon further heating the iron carbonate decomposes into iron oxide and carbon dioxide.

▲   As the contents, iron(II), pours from the tube, it comes in contact with the oxygen in the air and is oxidized exothermically to iron(III).

▲   To show students that ligand strength is real, repeat the procedure with $FeCl_2$. The oxalate will not replace the chloride, and the yellow oxalate is not produced. If you heat this product to black, it will not oxidize when poured from the test tube.

Reactions

Materials

▲ Iron(II) sulfate heptahydrate, $FeSO_4 \cdot 7H_2O$, crystals.

▲ Oxalic acid, $C_2H_2O_4$, crystals.

▲ Nickel(II) sulfate hexahydrate, $NiSO_4 \cdot 6H_2O$, crystals.

▲ Cobalt(II) nitrate hexahydrate, $Co(NO_3)_2 \cdot 6H_2O$, crystals.

▲ Manganese(II) sulfate monohydrate, $MnSO_4 \cdot H_2O$, crystals.

1.   Iron:

$$Fe^{2+}(aq) + H_2C_2O_4(aq) + 2H_2O(l) \rightarrow$$
$$FeC_2O_4 \cdot 2H_2O(s) + 2H^+(aq)$$

$$FeC_2O_4 \cdot 2H_2O(s) \xrightarrow{\Delta}$$
$$FeO(s) + 2H_2O(g) + CO(g) + CO_2(g)$$

$$4FeO(s) + O_2(g) \rightarrow 2Fe_2O_3(s) + heat$$

2.   Cobalt:

$$Co^{2+}(aq) + H_2C_2O_4(aq) \rightarrow$$
$$CoC_2O_4 \cdot 2H_2O(s) + 2H^+(aq)$$

$$CoC_2O_4 \cdot 2H_2O(s) \xrightarrow{\Delta}$$
$$CoO(s) + 2H_2O(g) + CO(g) + CO_2(g)$$

$$4CoO(s) + O_2(g) \rightarrow 2Co_2O_3(s) + heat$$

3.   Nickel:

$$Ni^{2+}(aq) + H_2C_2O_4(aq) \rightarrow$$
$$NiC_2O_4 \cdot 2H_2O(s) + 2H^+(aq)$$

$$NiC_2O_4 \cdot 2H_2O(s) \xrightarrow{\Delta}$$
$$NiO(s) + 2H_2O(g) + CO(g) + CO_2(g)$$

$$4NiO(s) + O_2(g) \rightarrow 2Ni_2O_3(s) + heat$$

4.   Manganese:

$$Mn^{2+}(aq) + H_2C_2O_4(aq) \rightarrow$$
$$MnC_2O_4 \cdot 2H_2O(s) + 2H^+(aq)$$

$$MnC_2O_4 \cdot 2H_2O(s) \xrightarrow{\Delta}$$
$$MnO(s) + 2H_2O(g) + CO(g) + CO_2(g)$$

$$4MnO(s) + O_2(g) \rightarrow 2Mn_2O_3(s) + heat$$

Notes

1.   The molecular weight of iron(II) sulfate hepta-hydrate is 278.03 g/mol, and that of oxalic acid is 126.07 g/mol. About 0.035 mol of each produces about 5 g of the oxalate. A heaping spoonful of the oxalate placed into a large test tube and heated is enough to produce a dramatic effect seen by an average-sized class.

2.   As the oxalic acid reacts with the salt, the color changes from the color of the solution to the color of the insoluble oxalate.

3.   The quantities suggested under Procedure allow for a slight excess of oxalic acid.

4.   Washing the precipitate well is very important, as contaminants tend to reduce the effectiveness.

References

The Merck Index, 10th ed.; Merck: Rahway, NJ, 1983; p 250.

Miller, C. D. *J. Chem. Educ.* **1987**, *64*, 545.

Newth, G. S. *Chemical Lecture Experiments;* Longmans, Green, & Co.: New York, 1892; p 107.

Roussin's Black Salt

When colorless nitric oxide gas is passed through a colorless solution of iron(II) sulfate–potassium thiosulfate, an intensely black suspension is produced.

Procedure

Please consult the Safety Information before proceeding. Perform this demonstration in a fume hood.

1. Stopper a 250-mL Erlenmeyer flask with a two-holed rubber stopper fitted with a long-stem funnel in one hole and glass bend and delivery tube in the other hole. When ready, place 10 g of copper turnings in the flask and stopper. Carefully add 50 mL of 6 M nitric acid through the funnel. Wait for 1 min to dispel any NO_2 gas.

2. After 1 min place the end of the delivery tube in a test tube containing 10 mL of iron(II) sulfate–potassium thiosulfate solution. Observe

Safety Information

1. Concentrated nitric acid causes severe burns, and contact with the eyes may result in a total loss of vision. Inhalation of the vapors may cause coughing and choking. Inflammation and ulceration of the respiratory tract may occur.

2. Sodium thiosulfate is a strong reducing agent, and contact with the skin may be irritating.

3. Iron sulfate-thiosulfate may be harmful by ingestion, inhalation, or skin absorption. →

the dense black precipitate forming. After about 30 s remove the delivery tube. Carefully unstopper the flask and fill the flask with water to stop the reaction.

3. The precipitate can be decanted or filtered and dried. Observe the shiny black crystals under a microscope.

Concepts

▲ When nitric oxide gas is added to a solution of iron(II) sulfate saturated with potassium thiosulfate, the iron forms a pseudotetrahedral shape. The nitric oxide acts as an oxidizing agent and oxidizes the iron(II) to iron(IV); thus the unpaired electron on the nitric oxide gains this electron to form a bonding pair. This shape is formed by one of the sulfur atoms in each of the thiosulfate ions bridging with the two Fe(IV) ions. This salt was first characterized by Paul Roussin in 1878.

▲ The observed structure indicates a formal charge of 4+ for the iron and some possible Fe–Fe interaction. Roussin produced at least one other complex iron salt (Roussin's red salt) with a different structure, which we have not been able to synthesize as conveniently.

▲ The salt forms an interesting tetrahedral complex for iron, and much research has been done concerning the potential use of this complex as a homogeneous catalyst.

▲ The structure of Roussin's salt is

Reactions

1. $3Cu(s) + 8HNO_3(aq) \rightarrow$

$$3Cu(NO_3)_2 + 2NO(g) + 4H_2O(l)$$

2. $2K^+(aq) + 2Fe^{2+}(aq) + 4NO(g) + 2S_2O_3^{2-}(aq) \rightarrow$

$$K_2[Fe_2(NO)_4(S_2O_3)_2](s)$$

Notes

1.   As the iron sulfate–potassium thiosulfate solution sits, it become slightly brownish-orange tinged. This color change does not affect the reaction. This solution, once prepared, can be saved.

2.   The nitric oxide is easily and safely prepared by Step 1 of the Procedure. A lecture bottle of gas is extremely convenient.

3.   After adding the nitric acid to the copper turnings, it is necessary to wait 1 min or so to dispel brown NO_2 and oxygen gases from the flask. This will allow relatively pure NO gas to be expelled into the test tube.

4.   Roussin's red salt is found in Chinese cabbage and can be isolated.

Materials

▲   Copper turnings.

▲   Potassium thiosulfate, saturated $K_2S_2O_3$: Dissolve 50 g in 50 mL of distilled water.

▲   Iron(II) sulfate heptahydrate, 0.1 M $FeSO_4 \cdot 7H_2O$: Dissolve 0.77 g in the 50 mL of potassium thiosulfate solution.

▲   Nitric acid, 6.0 M HNO_3: Carefully add 38.2 mL of concentrated acid to enough water to make 100 mL of solution.

References

Partington, J. R. *A Textbook of Inorganic Chemistry;* Macmillan: London, 1950; pp 938–939.

Greenwood, N. N.; Earnshaw, A. *Chemistry of the Elements;* Pergamon: New York, 1984; pp 514, 1272.

Sulfur Black

When tannic acid, sulfur, and sodium carbonate are heated in a test tube, an intensely black dye is produced.

Procedure

Please consult the Safety Information before proceeding.

1. Place 2 g each of sulfur, tannic acid, and sodium carbonate into a small test tube and heat the test tube carefully in a hood or behind a safety shield. The mixture will ignite, and vapor will be expelled from the mouth of the tube.

2. Cool the test tube to room temperature, and then add 5 mL of water. Stir the mixture thoroughly with a stirring rod, breaking up any chunks. Observe the intense black color.

Safety Information

1. Tannic acid is toxic when ingested and considered a carcinogen. Although tannic acid has a medical history as an astringent and antidote for heavy metal poisoning, extreme care should be used when handling. Avoid contact with the mucous membranes and skin, and wear gloves when handling.

2. Sulfur is toxic when ingested or when the dust is inhaled. Avoid contact with the mucous membranes and skin, and wear gloves when handling.

3. Sodium carbonate decahydrate is nonhazardous, but prudent laboratory practices should be exercised.

Materials

▲ Sulfur, S_8, powder.

▲ Tannic acid, powder.

▲ Sodium carbonate deca-hydrate, $Na_2Co_3 \cdot 10H_2O$, powder.

Concepts

The three substances are fused together to produce a primitive dye. We could not find information on the composition of this substance. The chemistry of tannins is very complex and difficult; the composition and structure are disputed. *The Merck Index* gives the empirical formula of corilagin, which is one of the tannins, as $C_{27}H_{24}O_{18}$. The commercial compound is given as $C_{76}H_{52}O_{46}$. It appears from the reaction, with little sulfur odor, that sulfur reacts with the tannic acid, which is an ester of a sugar. Also, the role of the sodium carbonate may be to provide a basic environment and the cation for the final product, a sodium salt.

Reactions

$$S_8(s) + Na_2CO_3(s) + C_{76}H_{52}O_{46}(s) \rightarrow$$
$$\text{complex product of uncertain composition}$$

Notes

1. The reaction releases gas, probably carbon dioxide, and can sputter very hot solid from the mouth of the test tube. If you cannot do this reaction in a hood or behind a safety shield, be sure to point the test tube away from people.

2. A piece of cotton can be placed in the liquid, and the effect of the dark brown color will be shown better.

References

The Chemcraft Book; Porter Chemical: Hagerstown, MD, 1928; p 49.

The Merck Index, 9th ed.; Windholtz, M., Ed.; Merck: Rahway, NJ, 1976; p 1301.

Colored Bands

When a drop of a pale blue solution containing three different metal ions is placed on a piece of white filter paper saturated with iodide ions, three different-colored concentric rings appear.

Procedure

Please consult the Safety Information before proceeding.

1. Saturate two pieces of 12.5-cm filter paper with potassium iodide solution. Let them dry.

2. Place 1 drop of the first metal ion solution in the center of one piece of filter paper. Observe the colored bands that form.

3. Place 1 drop of the second metal ion solution in the center of the second piece of filter paper. Observe the slightly different colored bands.

5. Lead nitrate is toxic by ingestion, inhalation, or subcutaneous routes.

6. Mercury(II) nitrate monohydrate is toxic by ingestion, inhalation, or subcutaneous routes. Practice prudent laboratory procedures.

Concepts

▲ The filter paper is saturated with iodide ions, which readily react with each of the metal ions in a specific way. Mercury forms an intensely orange iodide. Lead forms a brilliant yellow iodide. Bismuth forms an intensely black iodide. Copper reacts with iodide to form copper(I) ions and liberates free iodine, which colors the paper brown.

▲ The different bands also strikingly show the different rates of capillary action for each ion. Copper has the fastest rate of diffusion, and mercury has the slowest. Lead and bismuth are about the same, and when used together, the black color of the bismuth iodide obscures the yellow of the lead iodide.

▲ The rate at which the products are formed is also shown by the order of the concentric bands. Although copper moves outward with the highest rate, it also reacts with the iodide to produce copper(I) ions and free iodine at a slower rate than lead or bismuth react to form their iodides.

Reactions

$$Bi^{3+}(aq) + 3I^-(aq) \rightarrow BiI_3(s)$$
$$Pb^{2+}(aq) + 2I^-(aq) \rightarrow PbI_2(s)$$
$$Hg^{2+}(aq) + 2I^-(aq) \rightarrow HgI_2(s)$$
$$2Cu^{2+}(aq) + 2I^-(aq) \rightarrow 2Cu^+(aq) + I_2(s)$$

Notes

1. The filter paper should be prepared just prior to use but long enough to be completely dry for the demonstration.

2. The concentration of the different ions in each mixed metal solution made may have to be adjusted slightly to obtain wide distinct bands that are easily observable. If a band is too faint or narrow, increase the amount of solution that causes that band, by drops. For instance, if the lead is not very obvious, to the original 3 mL add 1 drop and mix the solution. Repeat the procedures; if this color band is still not apparent, add additional drops until a reaction occurs with a wide distinct color band.

3. F. F. Runge (1850), as reported by Ihde (1984), wrote of the usefulness of the different rates of capillary action of dissolved substances on porous materials—namely, aqueous solutions on paper. However, it was not until 1903 (Mikhail Tswett) that paper chromatography—adsorption chemistry—came into its own with the successful separation of plant pigments. Tswett, during the years 1906–1914 as reported by Ihde (1984), used the extraction properties of petroleum ether and separated chlorophylls and associated xanthophylls and carotenes. Feigl is also given extensive credit for making use of adsorption properties in "qualitative spot tests".

References

Feigl, F. *Spot Tests, Vol. I, Inorganic Chemistry*, 4th English translation; Elsevier: Houston, TX, 1954; pp 71–73

Ihde, A. J. *The Development of Modern Chemistry;* Dover: New York, 1984; pp 570–580.

Materials

▲ Potassium iodide, 1.0 M KI: Dissolve 16.6 g in water to make 100 mL of solution.

▲ Copper(II) nitrate trihydrate, 0.1 M $Cu(NO_3)_2 \cdot 3H_2O$: Dissolve 0.25 g in water to make 10 mL of solution.

▲ Lead(II) nitrate, 0.1 M $Pb(NO_3)_2$: Dissolve 0.35 g in water to make 10 mL of solution.

▲ Bismuth nitrate pentahydrate, 0.1 M $Bi(NO_3)_3 \cdot 5H_2O$: Dissolve 0.50 g in water to make 10 mL of solution.

▲ Mercury(II) nitrate monohydrate, 0.1 M $Hg(NO_3)_2 \cdot H_2O$: Dissolve 0.27 g in water to make 10 mL of solution.

▲ First metal ion solution, Bi^{3+}, Pb^{2+}, and Cu^{2+}: Add 1 mL of bismuth nitrate, 0.5 mL of copper nitrate, and 3 mL of lead nitrate solutions to a test tube and mix thoroughly.

▲ Second metal ion solution, Bi^{3+}, Cu^{2+}, and Hg^{2+}: Add 1 mL of bismuth nitrate, 1 mL of mercury nitrate, and 0.5 mL of copper nitrate solutions to a test tube and mix thoroughly.

Synthesis of Eosin and Erythrosin Dyes from Fluorescein

Elemental bromine and iodine react with fluorescein to produce two new fluorescent dyes that are pH sensitive.

Procedure

Please consult the Safety Information before proceeding.

1. Place 100 mL of a fluorescein solution in a beaker and demonstrate its fluorescence with an ultraviolet light source. Save about 10 mL for use later.

2. Divide the remaining volume of the solution between two separate flasks. Add 10 mL of bromine water to one flask and 10 mL of iodine water to the other. Place the flasks on a magnetic stirrer and stir for 5 min. The color of the bromine flask will change from orange–green to reddish orange. The color of the iodine flask will change to a reddish yellow.

3. Extreme care must be exercised in the use of bromine water. Contact with the liquid will result in tissue damage. Inhalation of the vapor may cause serious lung damage; contact with the eyes may result in a total loss of vision. Work in a fume hood and wear gloves, a rubber apron, and goggles.

4. Fluorescein is toxic to all tissues. Inhalation of the dust or concentrated mist may cause damage to the respiratory tract. Wear gloves, an apron, and goggles when handling.

5. Concentrated ammonia, when inhaled, will cause edema of the respiratory tract, spasm of the glottis, and asphyxia. Treatment must be prompt to prevent death.

3. Decant 10 mL of each solution into two different test tubes. Check the pH of each solution. The eosin, tetrabromofluorescein, will be about 3.0, and the erythrosin, tetraiodofluorescein, will be about 4.0. Check the pH of the fluorescein solution as well. Record these data.

4. Observe and record the colors of the three solutions. Darken the room and place each in front of an ultraviolet light source. Observe the fluorescence and color of each.

5. Slowly, dropwise, add 1 M aqueous NH_3 to the tetrabromofluorescein solution until fluorescence begins. Record the pH.

6. To each test tube, slowly, dropwise, add 1 M hydrochloric acid until fluorescence is quenched. Record the pH of each solution. Repeat with 5 mL of the fluorescein solution from Step 1.

7. Add drops of aqueous ammonia until fluorescence begins with the fluorescein. Record the pH.

8. Take the remaining 5 mL of the fluorescein solution from Step 1 and add 5 mL of bleach. Slowly and carefully add 2 drops of concentrated HCl. The reaction produces Cl_2 gas. However, this halogen does not react with the fluorescein.

Concepts

▲ Fluorescence is due to the absorption of shorter-wavelength light (higher-energy photons) and the emission of longer-wavelength light (lower-energy photons) by a substance. The difference in energy of the photons is due to that lost in collisions within the system and results in increased kinetic energy or excited vibrational states.

▲ The bromine and iodine react with the fluorescein molecule to produce two new molecules: tetrabromofluorescein (eosin) and tetraiodofluorescein (erythrosin), respectively.

▲ All three indicators, fluorescein, eosin, and erythrosin, are pH sensitive and are used in titration when the color change of nonfluorescing indicator would be obscured by the color of the solutions.

▲ This procedure was a classic method for the detection of trace amounts of bromine and iodine. According to Frehden and Huang (1939), as reported in Feigl (1954), as little as 6 parts per million of iodine will cause a color change in fluorescein.

Reactions

fluorescein $+ \ 4Br_2 \longrightarrow$ tetrabromofluorescein (eosin) $+ \ 4HBr$

fluorescein $+ \ 4I_2 \longrightarrow$ tetraiodofluorescein (erythrosin) $+ \ 4HI$

Notes

1. The bromine water can be produced several ways, but dissolving pyridinium bromide perbromide in water is by far the safest. The addition of commercial bleach to sodium or potassium bromide with the addition of a little HCl also will produce bromine water. A similar procedure will work for iodine by the addition of bleach and HCl to sodium or potassium iodide. Because the chlorine vapor produced will not react with the fluorescein, this procedure is a chemical alternative.

2. The pH range for fluorescence is rather narrow, so care must be taken when adding the drops of acid or base.

Materials

▲   Fluorescein: Add 0.1 g to 100 mL of distilled water.

▲   Hydrochloric acid, 1 M HCl: Add 8.3 mL of concentrated HCl to enough water to make 100 mL of solution.

▲   Aqueous ammonia, 1 M NH_3: Add 6.8 mL of concentrated aqueous ammonia to enough water to make 100 mL of solution.

▲   Bromine water, saturated Br_2: Place 1 g of pyridinium bromide perbromide in 10 mL of water in a test tube. Wait about 10 min for the solution to become saturated.

▲   Iodine water, saturated I_2: Add 0.5 g of iodine crystals to 10 mL of distilled water and let the mixture sit overnight.

▲   Bleach: commercial laundry product, about 5%.

3.   You can compare the fluorescence of the compounds produced with that of commercially prepared eosin and erythrosin.

References

Feigl, F. *Spot Tests, Vol. I, Inorganic Chemistry*, 4th English translation; Elsevier: Houston, TX, 1954; pp 71–73

Nassau, K. *The Physics and Chemistry of Color;* Wiley: New York, 1983.

Tomicek, O. *Chemical Indicators;* Butterworths: London, 1951.

Organic and Biological

Bleaching the Starch–Iodine Complex

Combining starch and iodine solutions results in the formation of a blue–black complex. Addition of bleach to the blue–black complex removes the color.

Procedure

Please consult the Safety Information before proceeding.

1. Pour about 15 mL of starch solution into a Petri dish and place the dish on an overhead projector.

2. Using a dropper, place about 4 drops of iodine solution in various spots in the starch solution, noting the blue-black color.

3. Using a different dropper, place drops of bleach on the blue–black spots to remove the color.

Materials

▲ Starch, $(C_6H_{10}O_5)_x$: Dissolve 1 g of starch in a few milliliters of water to make a smooth paste. Boil 200 mL of water and slowly, with stirring, add it to the starch. Cool the solution before using.

▲ Iodine–potassium iodide solution, I_2–KI: Dissolve 1.3 g of I_2 and 4.0 g KI in 50 mL of water. Dilute the solution to 1000 mL.

▲ Bleach: Any commercial bleach, about 5%, with no dilution.

Concepts

▲ Hypochlorous acid, HOCl, oxidizes some colored substances to colorless compounds.

▲ Chemical bleaching was introduced by C. L. Berthollet in 1785 when he recognized that solutions containing chloride had a strong decolorizing action on cloth. The fact that chlorine was generated from hydrochloric acid with manganese(IV) oxide kept bleach from general use. Charles Macintosh, who improved on a process begun by Charles Tennant, made a dry powder that was easily transported and could be turned into a bleaching agent by acidification with sulfuric acid. The industrial manufacture of bleach was begun at the St. Rollox Works, founded by Tennant in Glasgow in 1799.

▲ In 1814, the starch–iodine color reaction was described by J. J. Colin and H. F. Gaultier de Claubry.

▲ See the Concepts section of Investigation 52.

Reactions

$$OCl^-(aq) \rightarrow Cl^-(aq) + O\cdot(aq)$$

$O\cdot$ is an oxygen radical, which is responsible for oxidizing the colored dye to the colorless oxidized form.

Notes

1. The reaction could also be done in a test tube or beaker but is more visible on an overhead projector.

2. Place several drops of iodine at different locations on a piece of paper to show students the blue–black color that results from the presence of starch.

References

Greenwood, N. N.; Earnshaw, A. *Chemistry of the Elements;* Pergamon: New York, 1984; p 921.

Ihde, A. J. *The Development of Modern Chemistry;* Dover: New York, 1984; p 449.

Kemp, D. S.; Vellaccio, F. *Organic Chemistry;* Worth: New York, 1980; p 993.

Nebergall, W. H.; Schmidt, F. C.; Holtzclaw, H. F., Jr. *College Chemistry;* Heath: Lexington, MA, 1976; p 505.

Newth, G. S. *Chemical Lecture Experiments;* Longmans, Green, & Co.: New York, 1892; p 107.

Teitelbaum, R. C.; Ruby, S. L.; Marks, T. J. *J. Am. Chem. Soc.* **1980**, *102*(10), 3322.

Modified Starch versus Polystyrene

Modified starch (Eco-foam) and polystyrene are each placed in both water and acetone. Eco-foam dissolves only in water, and polystyrene dissolves only in acetone.

Procedure

Please consult the Safety Information before proceeding.

1. Place 50 mL of acetone in each of two 150-mL beakers.

2. Place 50 mL of water in each of two additional 150-mL beakers.

3. Drop one piece of polystyrene in each beaker of water and one piece in each beaker of acetone.

4. Repeat Steps 1–3 with modified starch (Eco-foam).

Safety Information

1. Acetone is a volatile, highly flammable liquid. It should not be used near any open flames. Also keep acetone away from plastic eyeglass frames, jewelry, pens and pencils, and rayon garments. Prolonged topical use may cause dryness, and inhalation may produce headaches and bronchial irritation.

2. Although starch is a food substance, no substance should be tasted in the laboratory.

Materials

▲ Acetone, $(CH_3)_2CO$.

▲ Polystyrene packing material.

▲ Eco-foam, modified starch. Samples can be acquired from American Excelsior Company, 850 Avenue H East, Arlington, TX (telephone: 817–640–1555).

Concepts

▲ Eco-foam is a starch-based material developed by National Starch and Chemical Company. It is made from a high-amylose cornstarch called Hylon VII (95%) modified by treatment with propylene oxide (5%). Moisture and the natural properties of Eco-foam are used to blow the material into its expanded, low-density condition. It became available in 1991 and is used as a packing agent.

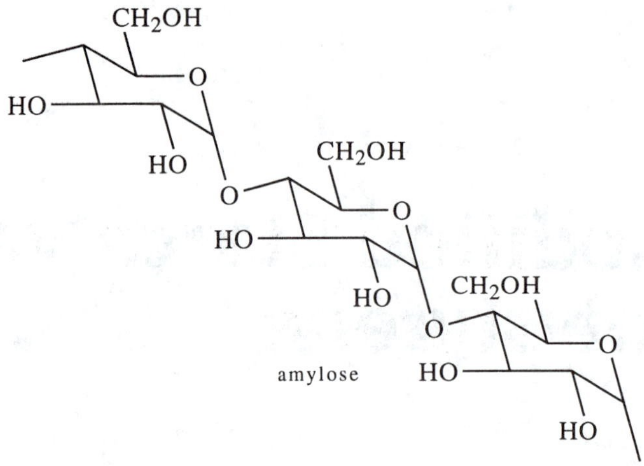

amylose

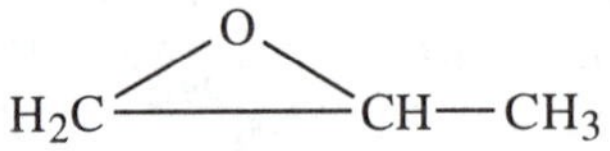

propylene oxide

▲ Propylene oxide treatment improves the cornstarch's degree of expansion, cell structure, resilience, and compressibility.

▲ Once dissolved in water, Eco-foam is 99% biodegradable in soil and soil-like environments.

▲ Polystyrene is made by polymerizing styrene. Styrene was first isolated in the 19th century. Early polymers were brittle and cracked easily. Polymerization of styrene did not become important until 1937, when a high-purity monomer was manufactured. After World War II, there was rapid growth in its use, especially in polystyrene foams.

polystyrene   polymer

▲ In the 1950s, foamable polystyrene beads were developed by BASF, and the blowing agents used to expand the polystyrene were pentane or hexane. The Dow Chemical Company developed an elongated shape of expandable polystyrene, which prevented items from settling in the shipping container.

▲ The choice of blowing agent determines the density and physical properties of polystyrene. Safety considerations such as flammability of hydrocarbons and environmental concerns about chlorofluorocarbons and their effect on the ozone layer must be considered when choosing the blowing agent.

Notes

1. Have your students look at and touch the two different substances and make observations about them. They should recognize the traditional "peanuts" but perhaps have not had an opportunity to see the new peanuts.

2. After the two have reacted with the appropriate solvent—polystyrene with acetone and Eco-foam with the water—show the students water so that they can see that the Eco-foam has dissolved. Remove the polystyrene from the water with tweezers, hold it up, pull it out into a string until it becomes brittle, wash it off, and allow the students to pass it around.

References

Capindale, J. *CHEM 13 News* **1991**, *204*, 8.

Kirk–Othmer Encyclopedia of Chemical Technology, 3rd ed.; Wiley: New York, 1983; Vol. 21, pp 770, 836, 840, 842.

Slowinski, E. J.; Wolsey, W. C.; Masterton, W. L. *Chemical Principles in the Laboratory*, 3rd ed.; Saunders: Philadelphia, PA, 1981; p 269.

Pyrolysis and Recrystallization

Sucrose and *p*-dichlorobenzene are both heated. One undergoes a chemical change, and the other a physical change.

Procedure

Please consult the Safety Information before proceeding.

1. Have your students make observations about the color and texture of table sugar (sucrose) and *p*-dichlorobenzene.

2. Place 10 mL of water in each of two separate test tubes. Add 1.0 g (1/2 tsp) of sugar to one tube and 1.0 g (1/2 tsp) of *p*-dichlorobenzene to the other. Stir the mixtures to dissolve. Have students observe the differences in dissolving.

3. Gently heat approximately 5 g of sugar in a Pyrex test tube for a couple of minutes, observing the many changes it undergoes.

Materials

▲   Table sugar (sucrose), $C_{12}H_{22}O_{11}$.

▲   *p*-Dichlorobenzene, $C_6H_4Cl_2$, crystals.

4.   Pass the test tube around after it cools.

5.   Place 10–12 crystals of *p*-dichlorobenzene in a small beaker on a hot plate. Place an evaporating dish full of ice on the top of the beaker.

6.   Gently heat the solid, allowing the students to observe its melting and crystallizing on the bottom of the evaporating dish, where long white crystals collect.

7.   Again place 10 mL of water in two separate test tubes. Compare the reaction of the carbon from the test tube with the crystals from the bottom of the evaporating dish.

Concepts

▲   Sucrose goes through a chemical change as it is decomposed or broken down by heat—pyrolysis—into various other substances, including liquids, solids, and gases. It is a chemical change, as evidenced by the difference in the starting material, a white crystalline solid, and the final product, a black brittle solid.

▲   *p*-Dichlorobenzene goes through physical changes. It remains the same chemical substance throughout, with changes only in its physical state. It changes from solid to liquid to gas to solid.

Physical Changes in State	Name
Solid to liquid	melting
Liquid to solid	freezing
Liquid to gas	vaporization
Gas to liquid	condensation
Solid to gas	sublimation
Gas to solid	deposition

▲   The idea for this demonstration was developed by A. F. Holleman in 1901.

Reactions

p-Dichlorobenzene, $C_6H_4Cl_2(s)$ → liquid → gas → solid

Sucrose, $C_{12}H_{22}O_{11}$: Sugar melts at 160 °C to form barley sugar. Upon further heating to 170–180 °C decomposition occurs, with darkening, loss of water, and formation of so-called caramel. It eventually forms brittle black carbon.

Notes

1. Setting up the apparatus as suggested for the *p*-dichlorobenzene allows you to safely use this substance. Leave the evaporating dish on top of the beaker when the students are observing the crystals so they will not be exposed to the vapors.

2. White crystalline sucrose changes to a thick liquid with a gold color, then an amber color, and finally a black color. The thick liquid caramelizes and eventually forms a black ashy solid. A white gas is given off during the heating, and a gold condensation forms on the sides of the test tube. Many or all of the observations can be made if the sugar is heated slowly and for a sufficient length of time.

3. When heating the sugar, use a test tube that you do not mind discarding. It will not be worth trying to clean.

4. Students will probably recognize the characteristic odor of *p*-dichlorobenzene from its use in mothballs.

References

Fowles, G. *Lecture Experiments in Chemistry;* Blakiston: Philadelphia, PA, 1937; p 21.

Kingzett, C. T. *Chemical Encyclopedia;* Van Nostrand: New York, 1928; pp 110, 692.

Test Your Turmeric

A yellow strip of filter paper soaked in turmeric turns orange when 1 drop of boric acid plus sulfuric acid is added, and the same spot turns blue-gray upon addition of sodium hydroxide solution.

Procedure

Please consult the Safety Information before proceeding.

1. Place a matchhead-sized piece of sodium borate decahydrate on a watch glass.

2. Add several drops of concentrated sulfuric acid. Mix the solution carefully.

3. Dilute the mixture with 7 drops of distilled water. Mix the solution.

4. Place 1 drop of the solution on a strip of turmeric paper, noting the orange color.

5. Add a drop of sodium hydroxide solution on

Safety Information

1. Turmeric, although used as a spice, should not be tasted when used in the laboratory.

2. Although sodium borate decahydrate is used as an ingredient for washing mucous membranes, ingestion of 5–10 g by young children can cause severe vomiting, diarrhea, shock, and death. →

the same spot, noting the blue-gray spot with purple edges.

Concepts

▲ *The Merck Index* lists the major coloring component of turmeric as curcumin. Its scientific name is 1,7-bis(4-hydroxy-3-methoxyphenyl)-1,6-heptadiene-3,5-dione, with a formula of $C_{21}H_{20}O_6$. Its structural formula is

3. Sulfuric acid is very corrosive. Handle with caution and avoid contact with skin because it produces severe burns. Inhalation of the concentrated vapor may cause serious lung damage. Work in a fume hood, and wear gloves, a rubber apron, and goggles.

4. Sodium hydroxide is corrosive to all tissues. Inhalation of the dust or concentrated mist may cause damage to the respiratory tract. Wear gloves, an apron, and goggles when handling.

▲ The yellow pigment of turmeric was isolated from the rhizomes of the tropical perennial *Curcuma longa* in 1815 by Vogel and Pelletier. The isolation is described in *Annalen der Chemie und Pharmacie* (1842). Vogel found that turmeric paper turns an intense orange–red using an ethanol solution of boric acid and blue from ammonia and other alkaline substances.

▲ A solution of boric acid is one of the weakest acids known. It does not taste sour and will not redden litmus paper. If sulfuric or hydrochloric acid is added to sodium borate decahydrate, in an amount equivalent to the sodium present, H_3BO_3 crystallizes from the solution upon cooling. An aqueous solution of H_3BO_3 has powerful germicidal properties and is often used as an eyewash. Sodium borate decahydrate is also used for laundry purposes because it has the properties of a mild alkali.

▲ H_3BO_3, also written as $B(OH)_3$, is moderately soluble in water. It acts as a Lewis acid toward the OH^- present. It does not increase H^+ concentration by giving up H^+, but by accepting OH^- from H_2O according to

$$B(OH)_3(aq) + H_2O(l) \leftrightarrows B(OH)_4^-(aq) + H^+(aq)$$

or

$$H_3BO_3(aq) + H_2O(l) \leftrightarrows H_4BO_4^-(aq) + H^+(aq)$$

▲ Turmeric has been used as an indicator to determine the end point in volumetric analyses. It changes from yellow to red–brown in the pH range 7.4–8.6 when there is 0.1% of the indicator in 50% alcohol.

▲ Turmeric is highly sensitive to weak acids. It is used for dyeing and in pharmacy and medicine for antiflammatory activity.

▲ Faraday (1822, republished 1932) tested the effects of various salts on turmeric paper.

Reactions

$$Na_2B_4O_7 \cdot 10H_2O(s) + H_2SO_4(aq) \rightarrow$$
$$4H_3BO_3(aq) + Na_2SO_4(aq) + 5H_2O(l)$$

Notes

1. Check the individual reactions of concentrated sulfuric acid, sodium hydroxide, water, and a solution of sodium borate on separate pieces of turmeric paper.

Substance	Color
H_2SO_4	no change
NaOH	deep orange
H_2O	no change
$Na_2B_4O_7$	deep orange, almost brown

2. You might want to try different concentrations of sodium hydroxide solution, alone or as a part of the reaction, to determine if it makes any difference in the color of the paper or the time it takes for the color change.

3. Also, try weaker concentrations of sulfuric acid, alone, to determine if there is any color change.

4. You could use turmeric paper in place of litmus paper for other experiments.

Materials

▲ Sodium borate decahydrate, $Na_2B_4O_7 \cdot 10H_2O$, powder: Using a mortar and pestle, grind up about 3.5 g (1 Tbsp).

▲ Turmeric paper: Obtain turmeric spice from the grocery store. Add 2 g to 30 mL of distilled water. Stir for a couple of minutes, even though it will be lumpy. Place a piece of filter paper in the "solution" for about 10–15 min. Remove the filter paper, rinse it with distilled water, and lay it aside to dry. Cut it into strips about 1 cm wide and store in a closed container until needed.

▲ Sodium hydroxide, NaOH: Dissolve 3 g with 100 mL of distilled water.

▲ Sulfuric acid, concentrated H_2SO_4: Commercial stock solution.

References

Ann. Chem. Pharm. **1842**, *44*, 297 (translated by L. Gelmacher, The Peddie School, Hightstown, NJ).

Faraday, M. *Faraday's Diary* **1932**, *1*, 70.

Kingzett, C. T. *Chemical Encyclopedia;* Van Nostrand: New York, 1928; p 770.

Kolthoff, I. M. *Acid–Base Indicators*, 4th ed.; Macmillan: 1937; p 148.

Leach, A. E. *J. Am. Chem. Soc.* **1904**, *26*, 1210.

The Merck Index, 10th ed.; Merck: Rahway, NJ, 1983; p 382.

Newth, G. S. *Chemical Lecture Experiments;* Longmans, Green, & Co.: New York, 1892; p 227.

Noyes, W. A. *A Textbook of Chemistry;* Holt: New York, 1916; pp 365–366.

Roughley, P. J.; Whiting, D. A. *J. Chem. Soc., Perkin Trans. 1* **1979**, *19–23*, 2379.

Whitten, K. W.; Gailey, K. D.; Davis, R. E. *General Chemistry*, 4th ed.; Saunders: Fort Worth, TX, 1992; p 940.

Vogel; Pelletier. *J. Pharm.* **1815**, *2, 50.*

Osmotic Pressure

Osmosis, the movement of water molecules from an area of higher concentration of water molecules to an area of lower concentration of water molecules across a semipermeable membrane, is demonstrated.

Procedure

Please consult the Safety Information before proceeding.

1. From a piece of dialysis tubing that is 2.5 cm wide, cut a piece 6 cm long. Soak it in water until it is soft (see the illustration on the next page for the setup of the demonstration).

2. When it is soft, cut open the dialysis tubing so it makes a rectangle about 5 cm wide by 6 cm long. Shape the wet piece of tubing around the end of a 45-cm piece of glass tube (13-mm inner diameter, 15-mm outer diameter). Let this tubing dry before proceeding.

Safety Information

Although pancake syrup is a food substance, no substance should ever be tasted in the laboratory.

Materials

▲   Pancake syrup: as purchased from the grocery store.

▲   Dialysis tubing, 2.5 cm wide.

▲   Tygon tubing, 12.7-mm inner diameter, 15.8-mm outer diameter.

▲   Glass tube, 13-mm inner diameter, 15-mm outer diameter.

3.   Cut a 40-cm piece of Tygon tubing (12.7-mm inner diameter, 15.8-mm outer diameter). Carefully place the Tygon tubing over the end of the dialysis tubing and piece of glass. Pour water into the Tygon tubing to make sure no water is leaking through the dialysis tubing because of a tear. Pour the water out.

4.   Over a sink, pour pancake syrup into the glass tube to about a 6-cm height. Clamp the glass tube and Tygon tubing in place. Use the warmth of your hands to speed up the movement of the syrup. Let the tube remain in place 10–15 min.

5.   Unclamp the glass tube and Tygon tubing and pour water into the Tygon tubing to about 6 cm from its end. Tilt the glass tube and Tygon tubing to remove any air bubbles close to the dialysis tubing.

6.   Clamp the Tygon tubing and glass tube in place and adjust them so the height of liquid is the same in both tubes. Mark the height of the liquids and note the time.

7.   Continue measurements about every 30 min, or as often as possible for 1 day. Leave the apparatus in place for several days.

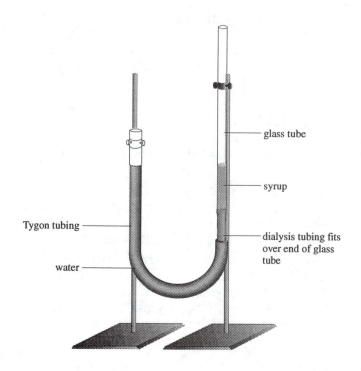

Concepts

▲ The process of osmosis occurs across a semi-permeable membrane, such as dialysis tubing, that permits some molecules to pass through but not others. It occurs because of a difference in the concentration of water molecules on the two sides of the membrane. Water moves from a higher concentration of water molecules to a lower concentration of water molecules.

▲ Because there is pure water on one side of the dialysis tubing—in the Tygon tubing—and a solution on the other side—in the glass tube—water moves from the Tygon tubing across the membrane and causes the height of the syrup (a solution) in the glass tube to increase because of osmotic pressure being exerted.

▲ Osmotic pressure can be calculated by using the following equation for dilute solutions:

$$\pi = MRT$$

where π is osmotic pressure, M is molarity, R is the gas constant, and T is temperature in kelvins. The syrup solution is not dilute; therefore, obtaining an accurate value for osmotic pressure is not possible.

▲ The density of the syrup in this demonstration in our lab was 1.345 g/mL. The syrup was made up of dextrose, maltose, polysaccharides, and water. The height of the syrup, because of osmotic pressure, could theoretically have reached around 300 m. However, as water moves into the syrup, it becomes diluted, the osmotic pressure is reduced, and the low pressure prevents the syrup from ever reaching this height.

▲ Eventually the height of the syrup no longer changes. At this point, a state of equilibrium has been reached, and a balance exists between the movement of water molecules across the membrane.

▲ At 15 °C, the osmotic pressure of seawater is 26 atm. The process of *reverse osmosis* is used to obtain fresh water from seawater. A pressure in excess of 26 atm is applied to seawater at 15 °C, and the tendency of the water to escape from the seawater exceeds that of pure water. A rigid semipermeable membrane separates pure water from the seawater.

Notes

1. Pour the syrup slowly so the height does not go above about 6 cm. If the height is too great, it may actually be too heavy, leak, and cause the dialysis tubing to slip off.

2. Do not warm the syrup in the glass tube with a Bunsen burner flame to speed up the movement of the syrup. The syrup will crystallize and go nowhere.

3. Other inner and outer diameters of Tygon tubing and glass tube could be used. The wider inner diameter of glass tube was chosen so it would be easier to pour syrup into the tube. No funnel is needed.

4. We tried using a sodium chloride solution in the glass tube and were unsuccessful, probably because of the low concentration, 0.01 M. A higher concentration of sodium chloride similar to that of seawater, 0.55 M, might work. We were successful using 3.0 g of polysucrose (molar mass of approximately 400,000 g/mol) gradually added, with stirring, to about 25 mL of water heated to boiling, cooled, and then poured into the glass tube. You might also find success using poly(vinyl alcohol) (molar mass of 125,000 g/mol). The idea behind the last two substances was to perhaps use the demonstration to determine the molar mass of a high molecular weight material.

5. Plastic wrap or a thin-weight plastic bag should work in place of dialysis tubing. The authors have substituted these in other demonstrations that called for dialysis tubing.

References

Campbell, A., The Torbitt and Castleman Co., Buckner, KY, July 1992, personal communication regarding syrup.

CRC Handbook of Chemistry and Physics, 66th ed.; Weast, R. C., Ed.; CRC: Boca Raton, FL, 1985.

Keeton, W. T. *Biological Science*, 3rd ed.; Norton: New York, 1980; pp 87–90.

McQuarrie, D. A.; Rock, P. A. *General Chemistry;* Freeman: New York, 1987; pp 392–395.

Petrucci, R. H. *General Chemistry*, 5th ed.; Macmillan: New York, 1989; pp 458–461.

Benedict's Solution and Cabbage Juice

When purple cabbage juice solution or a colorless sugar solution is heated with Benedict's solution, an amber–brown color results.

Procedure

Please consult the Safety Information before proceeding.

1. Place four test tubes, each containing 5 mL of Benedict's solution, in a boiling water bath. Heat the tubes for 1 min.

2. Add 1 pipetful of each of the three sugar solutions to three separate test tubes. Swirl the solutions to mix.

3. To the fourth test tube, add 5 mL of cabbage juice. Swirl the solution to mix.

Safety Information

1. Although the sugars used are food substances, no substance should ever be tasted in the laboratory.

2. Benedict's solution contains sodium citrate, sodium carbonate, and copper(II) sulfate. Copper(II) sulfate pentahydrate is a skin and respiratory irritant and is toxic by ingestion and inhalation. Sodium carbonate may be a skin irritant. Sodium citrate is poisonous by intravenous route. It is moderately toxic by subcutaneous route and mildly toxic by ingestion.

Materials

▲   Cabbage juice: Cut up a red cabbage into small pieces and boil in distilled water for 1 h. Pour off the purple solution and use it for the demonstration.

▲   Sugar solutions, 2% $C_6H_{12}O_6$: Dissolve 1.0 g in 50 mL of water in separate flasks. The sugars are glucose, galactose, and fructose.

▲   Benedict's solution: Order from a commercial source or prepare as follows. Dissolve 20 g of sodium citrate and 11.5 g anhydrous sodium carbonate in 100 mL of hot water in a 400-mL beaker. With continuous stirring, add slowly a solution of 2 g of copper sulfate pentahydrate in 20 mL of water. If the solution is not clear, filter it.

4.   Heat the four test tubes in the boiling water bath for 10 min, making observations.

Concepts

▲   Benedict's solution is a basic solution of Cu^{2+}. When heated with a reducing agent such as glucose, copper(I) oxide, Cu_2O, precipitates. Copper(I) oxide forms as a reddish-brown solid.

▲   In Benedict's solution the copper(II) ion is complexed with the citrate ion. The complex prevents precipitation of copper(II) hydroxide or copper(II) oxide.

▲   Simple sugars, or monosaccharides, can act as reducing agents for Benedict's solution. Examples include glucose, galactose, and fructose. These sugars have the formula $C_6H_{12}O_6$.

▲   Anthocyanin pigments are present in red cabbage in the form of glucosides. By hydrolysis, or the addition of water, anthocyanins are converted into glucose or other monosaccharides and colored anthocyanidins. The three parent anthocyanidins are named pelargonidin, cyanidin, and delphinidin.

▲   The reaction using Benedict's solution has been used to detect the presence of reducing sugars in the urine of diabetics. Urine does not normally contain glucose, but when a person is diabetic, glucose is present.

Reactions

1.   General reaction; CuO represents Benedict's solution:

$$RCHO(aq) + 2CuO(aq) + NaOH(aq) \rightarrow$$
$$RCOONa(aq) + Cu_2O(s) + H_2O(l)$$

Specific reaction using glucose as the reducing sugar:

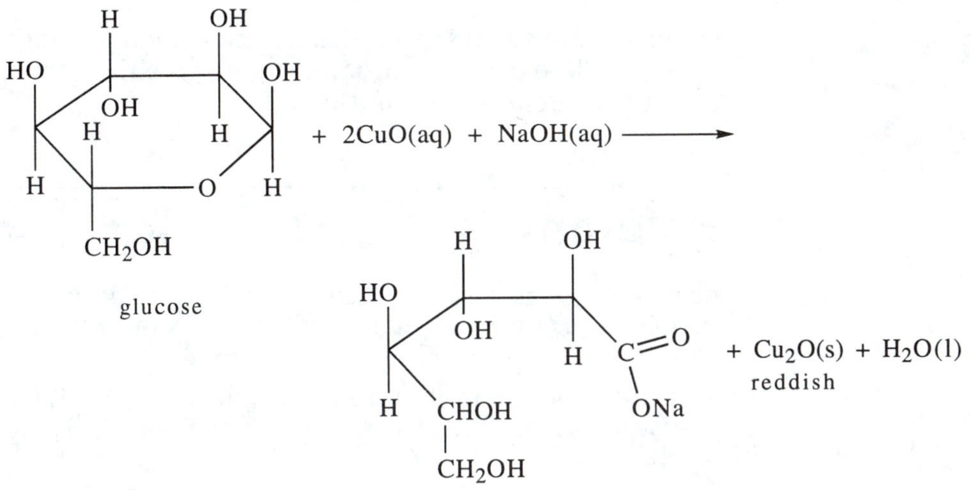

glucose

$+ \; 2CuO(aq) \; + \; NaOH(aq) \longrightarrow$

$+ \; Cu_2O(s) \; + \; H_2O(l)$
reddish

2. The hydrolysis reaction of cyanin, one of the three anthocyanin pigments present in red cabbage, to form cyanidin chloride and glucose is

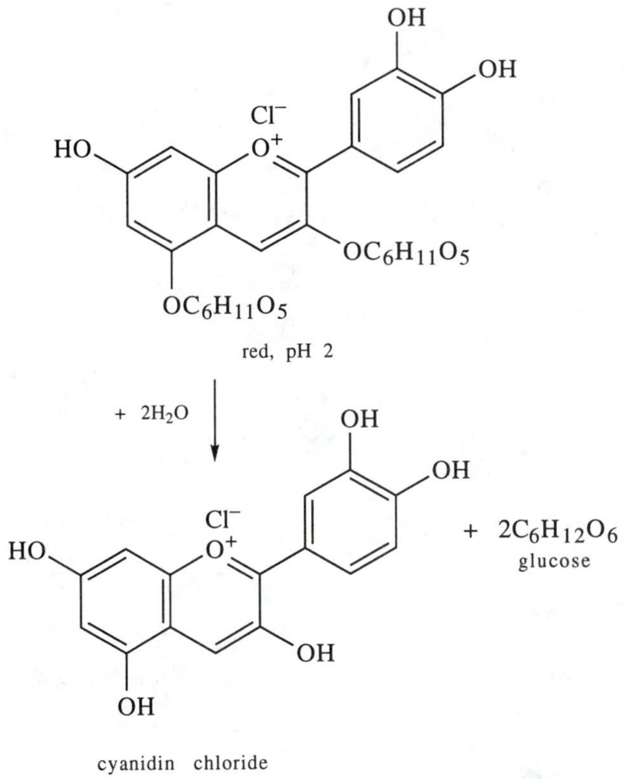

red, pH 2

$+ \; 2H_2O$

$+ \; 2C_6H_{12}O_6$
glucose

cyanidin chloride

Notes

Other simple sugars, or a disaccharide such as maltose, could be used, but not all sugars give the same result with Benedict's solution.

References

Adams, R.; Johnson, J. R.; Wilcox, C. F., Jr. *Laboratory Experiments in Organic Chemistry*, 5th ed.; Macmillan: New York, 1963; pp 208, 267.

Fuson, R. C.; Connor, R.; Price, C. C.; Snyder, H. R. *Brief Course in Organic Chemistry;* Wiley: New York, 1941; pp 39, 220.

Hollum, J. R. *Elements of General and Biological Chemistry*, 6th ed.; Wiley: New York, 1983; p 226.

Nebergall, W. H.; Schmidt, F. C.; Holtzclaw, H. F., Jr. *College Chemistry*, 5th ed.; Heath: Lexington, MA, 1976; p 869.

Purple Bottle

A colorless alkaline glucose solution containing two indicators is shaken. At first, it changes to a bright pink, and with further shaking to a bright purple. Upon sitting, the solution becomes pink and then colorless.

Procedure

Please consult the Safety Information before proceeding.

1. Place 50 mL of glucose solution and 50 mL of sodium hydroxide solution in a 250-mL bottle.

2. Add 2 drops of methylene blue indicator and an amount of Safarine-O powdered indicator about half the size of a pinhead to the bottle and gently swirl the solution.

Materials

▲ Sodium hydroxide, 1.0 M NaOH: Dissolve 10 g of NaOH in enough water to make 250 mL of solution.

▲ Glucose, 0.25 M $C_6H_{12}O_6$: Dissolve 11.3 g of $C_6H_{12}O_6$ in enough water to make 250 mL of solution.

▲ Methylene blue: Dissolve 0.001 g of methylene blue in 9 mL of 0.01 M NaOH and dilute to 100 mL with water.

▲ Safarine-O powdered indicator.

3. Stopper the bottle and let the solution remain undisturbed until it is colorless. A total of 10–15 min is required for the solution to become pink and then finally colorless.

4. Shake the bottle gently until the solution turns pink, and then shake the bottle violently until the solution turns purple.

5. Let the solution stand undisturbed until the color disappears.

6. Repeat Steps 4 and 5.

7. When the solution no longer changes color, open the bottle and let fresh air into the bottle. Repeat the shaking.

Concepts

▲ Glucose is a reducing sugar and reduces the two indicators to colorless forms.

▲ When the solution is shaken, the oxygen in the bottle above the solution dissolves and oxidizes the Safarine-O to the pink form.

▲ Upon further shaking of the solution, the methylene blue is also oxidized to the blue form.

▲ When the solution containing the oxidized forms is allowed to remain undisturbed, the sugar reduces the indicators back to their colorless forms (methylene blue is first to be reduced to the colorless form, which in turn keeps the Safarine-O in the oxidized form).

▲ When the solution is gently agitated, the Safarine-O is more easily oxidized, and only its oxidized form is visible. When the solution is further shaken, the colorless form of methylene blue is oxidized to the blue form.

▲ Upon sitting, the methylene blue is reduced back to the colorless form first, and then the Safarine-O is reduced back to the colorless form. However, because oxygen is in contact with and dissolves at the surface of the solution, a thin purple layer is always present.

Reactions

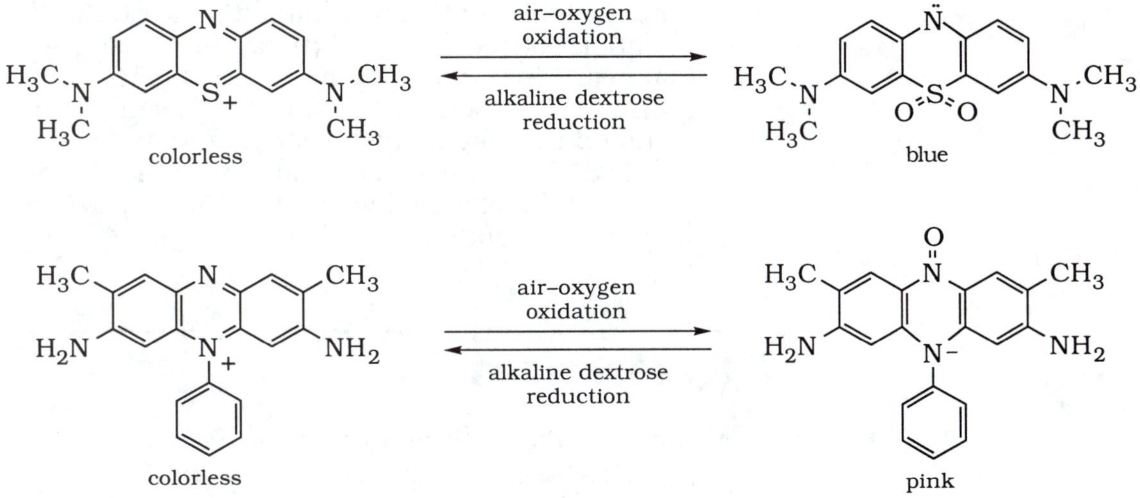

air–oxygen oxidation

alkaline dextrose reduction

colorless / blue

air–oxygen oxidation

alkaline dextrose reduction

colorless / pink

Notes

1. The two indicator solutions can be prepared well in advance and stored for the school year.

2. Add the two solutions together and add the two indicators (Steps 1–3 of the Procedure) and let sit for 10–15 min prior to presenting the demonstration. This initial reducing time will be longer.

3. When the solution no longer changes color, open the bottle to renew the oxygen in the bottle. You can blow into the bottle to help the students discover what is happening.

4. This is a variation of "The Blue Bottle Reaction" and "A Traffic Light Reaction" (Investigations 59 and 50, respectively, in Summerlin and Ealy's *Chemical Demonstrations*.)

5. The thin purple surface layer should lead to questions and the answer.

6. About 150 years ago a French immigrant, Martin Fugate, married American Elizabeth Smith; both carried a recessive gene. Their children possessed the gene combination that caused a reduction in the amount of the enzyme diaphorase. In the body, pyridine nucleotides reduce the hemoglobin to the colorless form. Then the methemoglobin oxidizes it back to the blue form. Without the needed amount of diaphorase, the reduction of methemoglobin back to hemoglobin is severely inhibited. The excessive amount of methemoglobin causes the skin to appear blue; thus they had "blue blood". The discovering scientist, Madison Cawein, concluded that the daily ingestion of a methylene blue tablet would reverse the process and lower the amount of methemoglobin, thus they would have red blood—pink skin tone.

References

Abbott, G. E. *Proc. Arkansas Acad. Sci.* **1947**, *2*, 45.

Abbott, G. E. *J. Chem. Educ.* **1948**, *25*, 100.

Campbell, F. B. *J. Chem. Educ.* **1963**, *40*, 578.

Color Index, 3rd ed.; Society of Dyers and Colourists; **1971**, *4*, 4451.

Lecture Demonstrations in General Chemistry, 1st ed.; Arthur, P., Ed. McGraw Hill: New York, 1939.

Mack, R. *N.C. Med. J.* **1982**, *4*, 292.

Mansouri, A. *Am. J. Med. Sci.* **1985**, *289*, 200.

The Merck Index, 9th ed.; Windholtz, M., Ed.; Merck: Rahway, NJ, 1976; p 791.

Olson, E. S. *J. Chem. Educ.* **1977**, *54*, 366.

Olson, E. S. *J. Chem. Educ.* **1983**, *60*, 493.

Summerlin, L. R.; Ealy, James L., Jr. *Chemical Demonstrations: A Sourcebook for Teachers;* American Chemical Society: Washington, DC, 1985.

Witt, O. N. *Ber. Dtsch. Chem. Ges.* **1886**, *19*, 3121.

Clear Bottle

When a bright blue, alcoholic hydroxide solution containing an indicator and zinc is shaken, the solution suddenly becomes colorless. Upon standing, the solution changes back to a bright blue color.

Procedure

Please consult the Safety Information before proceeding.

1. Place 150 mL of a 0.1% ethanol solution of thioxanthone sulfone into a 300-mL bottle.

2. Add about 15 g of zinc amalgam to the bottle. Wait until the solution turns blue.

3. Add 37 mL of a 5 M sodium hydroxide solution.

4. Stopper the bottle and shake it until the solution is colorless.

5. Allow the bottle to sit until the blue returns. Shake the bottle again.

Safety Information

1. Sodium hydroxide is corrosive to all tissues. Inhalation of the dust or concentrated mist may cause damage to the respiratory tract. It generates considerable heat upon forming an aqueous solution.

2. Ethanol is a volatile, highly flammable liquid. It should not be used near any open flames. Prolonged topical use may cause dryness, and inhalation may produce headaches. →

393

3. Nitric acid is corrosive to all body tissues. Inhalation of the concentrated vapor may cause serious lung damage; contact with the eyes may result in a total loss of vision. Continued exposure to the vapor of nitric acid may cause chronic bronchitis.

4. Glacial acetic acid is corrosive to all body tissues. Inhalation of the concentrated vapor may cause serious lung damage; contact with the eyes may result in a total loss of vision. Continued exposure to the vapor of acetic acid may cause chronic bronchitis.

5. Hydrogen peroxide, 30%, is a strong oxidant and should be used with caution. Avoid all contact with the skin and eyes.

6. Mercury(II) chloride is very toxic; use gloves and wash your hands very carefully after use.

Concepts

▲ The thioxanthone sulfone is reduced to the blue semiquinone-like free radical by the zinc amalgam in the alcoholic sodium hydroxide solution.

▲ The dissolved oxygen in the alcoholic alkaline solution oxidizes the semiquinone back to thioxanthone sulfone.

▲ After a few trials, the colorless portion of the reaction may begin to appear pink. This color is caused by the production of a stable red form.

▲ The color of the semiquinone is due to the charge associated with the C–O and the O–S=O ends of the center ring. Because of the reduction by zinc metal in a basic alcoholic medium, the electron transfers across the ring structure to the carbon–oxygen bond. This charge distribution and structure, which are analogous to those of the anthraquinone–anthrahydroquinone system, play an important role in the absorption in the yellow range and thus produce the blue color of the semiquinone.

Reactions

1.

colorless

blue

red (pink solution)

2.

blue → colorless

3. Reaction for the production of thioxanthone sulfone:

thioxanthen-9-one

$+ 2H_2O_2 \xrightarrow{\text{acetic acid}}$

thioxanthone sulfone

Materials

▲ Thioxanthone sulfone, $C_{13}H_{10}O_3S$: Place 5 g of thioxanthen-9-one in a round-bottom flask. Add 12 mL of 30% hydrogen peroxide and 45 mL of glacial acetic acid. Reflux the mixture with a magnetic stirrer for 1 h. Add 12 mL of water. Reheat the mixture to dissolve the precipitate. Cool the solution overnight, filter it, and wash it several times with a 50% aqueous solution of acetic acid. Add 10 mL of water, reheat the mixture to dissolve and recrystallize the precipitate. Repeat the washing. Dry the precipitate in an oven at a low temperature (75 °C) overnight. Add 1.0 g of the pale yellow precipitate to 1.0 L of 95% ethanol.

▲ Sodium hydroxide, 5 M NaOH: Add 100 g to enough water to make 500 mL of solution.

▲ Zinc amalgam: Add 10 g of granular zinc to 200 mL of water. Add 0.5 g of mercury(II) chloride. Add 2 mL of concentrated nitric acid. Shake the solution and wait about 5 min. Decant the solution and wash the amalgam with distilled water. Store under distilled water in a stoppered bottle.

Notes

1. The time it takes for the bottle to turn blue is dependent upon the purity of the thioxanthone sulfone. The time may vary from a few minutes to several hours—the higher the purity, the less time.

2. The blue–colorless reaction will continue for several days. Zinc amalgam (large mesh) sinks to the bottom quickly, produces a colorless solution, and is better than pure granular zinc or zinc dust.

3. This demonstration should be used in conjunction with "A Traffic Light Reaction" (Demonstration 50 in Summerlin and Ealy's *Chemical Demonstrations*) and with the comparison of the reducing rates of simple sugars (Investigation 68) and the "Purple Bottle" (Investigation 97) in this volume. All four bottles can be displayed at once or they can be shown in order: "Blue Bottle", "Purple Bottle", "Traffic Light", and "Clear Bottle" for the greatest educational value.

References

Abbott, C. E. *J. Chem. Educ.* **1948**, *25*, 100.

Castrillon, J. *J. Chem. Educ.* **1987**, *64*, 352.

Davis, E. G.; Smiles, S. J. *J. Am. Chem. Soc.* **1910**, *32*, 1290.

Davis, E. G.; Smiles, S. J. *J. Am. Chem. Soc.* **1911**, *32*, 640.

Fehnel, A. C. *J. Am. Chem. Soc.* **1949**, *71*, 1063.

Heymann, H. J. *J. Am. Chem. Soc.* **1949**, *71*, 260.

Nassau, K. *The Physics and Chemistry of Color;* Wiley-Interscience: New York, 1983; pp 140–150.

Summerlin, L. R.; Ealy, James L., Jr. *Chemical Demonstrations: A Sourcebook for Teachers;* American Chemical Society: Washington, DC, 1985; p 79.

Ullmann, F. von G., *Ber. Dtsch. Chem. Ges.* **1916**, *49*, 2487.

Halogenation of a Triphenylmethane Dye

When a colorless solution of metabisulfite–sulfite is added to a red fuchsin (rosaniline) solution, it becomes colorless. Adding glyoxal causes the color to reappear. Adding bromine water to the colorless solution causes it to become blue.

Procedure

Please consult the Safety Information before proceeding.

1. Place 1 mL of fuchsin (rosaniline) indicator solution in a large test tube. Add 4 mL of distilled water.

2. Add several drops of sodium metabisulfite–sodium sulfite solution and shake. If necessary, dropwise, add more solution until it becomes colorless. Divide the solution into two portions in two test tubes.

3. To one test tube, add 2 drops of glyoxal solution. The color of the fuchsin indicator is restored.

Safety Information

1. With all elemental halogens, work in a fume hood, and wear goggles, gloves, and a rubber apron. Concentrated halogen vapor causes edema of the respiratory tract, spasm of the glottis, and asphyxia.

2. Extreme care must be exercised in the use of bromine water. Contact with the liquid will result in tissue damage. Inhalation of the vapor may cause serious lung damage; contact with the eyes may result in a total loss of vision. Work in a fume hood, and wear gloves, a rubber apron, and goggles. →

3. Sodium metabisulfite-sulfite and ethylenediamine-tetracetic acid (EDTA) may be harmful by ingestion, inhalation, or skin absorption.

4. Glyoxal, 40%, is moderately irritating to the skin and mucous membranes.

4. To the second test tube, add 5 mL of bromine water. Observe the deep blue solution.

5. Repeat Steps 1–4 with brilliant green or malachite green indicator.

Concepts

▲ The basic fuchsin is a triphenylmethane dye that reacts with the sulfurous acid, H_2SO_3, to produce the colorless form of the indicator from the red quinoidal form. The sulfurous acid attacks the double bond of the central carbon and the double bond of the $=NH_2$ group.

▲ The leuco form is susceptible to bromination. The dissolved bromine attacks the leuco form at the sites of the bisulfite, HSO_3^-, and produces the blue brominated form.

▲ The addition of an aldehyde, such as glyoxal, CHOCHO, to the leuco form of fuchsin restores the red quinoidal form.

▲ Brilliant green, also a triphenylmethane dye, is converted to the colorless leuco form by the sulfurous acid. However, the bromine does not produce a colored form in this instance. Hence, the decolorized solution of brilliant green does not show a visible reaction with the bromine water.

▲ This reaction was once a very sensitive "wet" qualitative test for bromine in the presence of other halogens, because chlorine and iodine do not react with fuchsin.

▲ Structures of brilliant green and malachite green are as follows:

brilliant green sulfate

malachite green oxalate

basic fuchsin (colored)

Reactions

1. $RCHO(aq) + HSO_3^-(aq) \rightarrow RCH(OH)SO_3^-(aq)$

2.

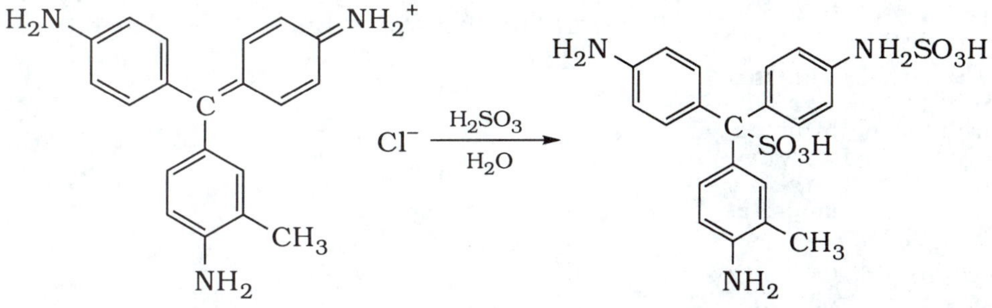

red   colorless (leuco base)

3.

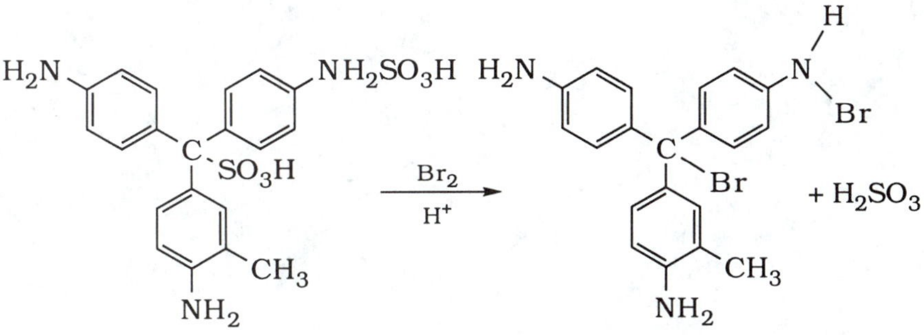

colorless (leuco base)   blue

Notes

1. See the Notes to Investigation 90 about producing bromine water.

2. The metabisulfite–sulfite solution is stabilized with EDTA and will last several weeks, at least.

3. A precipitate may form in the 40% glyoxal solution after prolonged storage, but it can be redissolved by warming to 50–60 °C. Work in an efficient fume hood when preparing solutions of this substance.

Materials

▲ Brilliant green, 0.1%: Dissolve 0.1 g of brilliant green in 100 mL of distilled water.

▲ Fuchsin, 0.1% $C_{20}H_{20}ClN_3$: Dissolve 0.1 g of fuchsin in 100 mL of distilled water.

▲ Bromine water, saturated Br_2: Add 1 g of pyridinium bromide perbromide, $C_5H_6Br_3N$, to 15 mL of water. Prepare at least 1 h before needed. It will remain saturated for a day.

▲ Glyoxal, 40% CHOCHO: commercial stock solution.

▲ Sodium metabisulfite–sodium sulfite, $Na_2S_2O_5$–Na_2SO_3: Dissolve 1.8 g of $Na_2S_2O_5$, 0.36 g of disodium ethylenediaminetetraacetate (Na_2 EDTA), and 0.3 g of Na_2SO_3 in enough distilled water to make 100 mL of solution.

▲ Malachite green, 0.1%: Dissolve 0.1 g of malachite green in 100 mL of distilled water.

References

Encyclopedia of Chemical Technology, 3rd ed.; Wiley-Interscience: New York, 1983; Vol. 23, p 408.

Feigl, F. *Spot Tests, Vol. I, Inorganic Chemistry*, 4th English translation; Elsevier: Houston, TX, 1954; p 287.

Bromine Rainbow in Tomato Juice

When a saturated solution of bromine water is added to tomato juice and stirred, a rainbow of colors is produced.

Procedure

Please consult the Safety Information before proceeding.

1. Place about 125 mL of tomato juice in a 250-mL graduated cylinder.

2. Fill a 125-mL pipet with bromine water. Carefully place the tip of the pipet near the bottom of the cylinder filled with tomato juice and add the bromine water.

Concepts

▲   The bromine attacks the double bonds in the conjugated system of the lycopene and the β-carotene in the tomato juice. The length of the conjugated system causes visible light in the blue range to be absorbed. This strong absorption produces the red color in tomato juice. The structure of lycopene is

Materials

▲ Bromine water, saturated: Add 5 g of pyridinium bromide perbromide to 250 mL of water in a loosely stoppered bottle and allow the solution to stand for about 1 h to attain saturation.

▲ Tomato juice: Purchase from the grocery store.

▲ When the double bonds are removed during the addition reaction, the length of the conjugated system shortens. Shorter length, in turn, causes the absorption spectrum to shift to the ultraviolet range and produces a colorless system.

▲ This attack is carried out by a *charge-transfer* complex on the olefin that produces brominium and bromide ions.

▲ This reaction is normally rapid, but because of the bilayer membranes (lipid aggregations), it is slower. The bromine is unable to get at the individual lipids. Hence, the ratio of concentrations of the bromine to lycopene must be very high (400:1).

▲ The green color is probably due to the combination of the blue color produced by the charge-transfer complex and the yellow of the bromine water.

Reactions

charge-transfer complex

Notes

1. The tomato juice should be at room temperature when it is added to the graduated cylinder.

2. See the Notes to Investigation 90 about producing bromine water.

3. By placing the tip of the pipet about 5 cm above the bottom of the graduated cylinder, you will still have some red (unreacted tomato juice) and then orange, yellow, green, and finally blue. This method will produce a rainbow of colors in the correct order.

5. Alternatively and easily, you can add the bromine to the top of the tomato juice, but this will produce an "incorrect" order of colors.

References

Banthorpe, D. V. *Chem. Rev.* **1970**, *70*, 306.

MacBeath, M. E.; Richardson, A. L. *J. Chem Educ.* **1986**, *63*, 1092–1094.

Olah, G. A.; Hockswender, T. R., Jr. *J. Am. Chem Soc.* **1974**, *96*, 3574.

Glyoxal–Resorcinol Polymer

Heating resorcinol in 40% glyoxal and then adding 1.0 M HCl causes formation of an orange–pink polymer.

Procedure

Please consult the Safety Information before proceeding.

1. Place 3.0 g of resorcinol and 4 mL of 40% glyoxal in a test tube.

2. Using a hot water bath, heat the contents of the test tube to dissolve the resorcinol and continue to heat to the boiling point.

3. Pour the test-tube contents into a disposable plastic beaker, add 1.0 mL of 1.0 M HCl, and stir the solution to mix.

4. The color will change from light yellow to orange to reddish. It will also start to boil and

Safety Information

1. Resorcinol is a skin and eye irritant and can be absorbed through the skin.

2. Glyoxal solution, 40%, is moderately irritating to the skin and mucous membranes. Work in an efficient fume hood when preparing solutions of this substance.

3. External contact with hydrochloric acid causes severe burns, and contact with the eyes may result in a total loss of vision. Inhalation causes coughing and choking with possible inflammation and ulceration of the respiratory tract. Work in a fume hood, and wear goggles, gloves, and a rubber apron.

Materials

▲   Hydrochloric acid, 1.0 M HCl: Adding acid to water, dilute 21.0 mL of concentrated acid with enough water to make 250 mL of solution.

▲   Glyoxal, 40% CHOCHO: commercial solution.

▲   Resorcinol, $C_6H_6O_2$: solid.

possibly burp, and a light orange–pink hard polymer will form.

5.   Wash the polymer thoroughly with water before handling.

Concepts

▲   A reaction occurs between phenol and formaldehyde. On the basis of a suggestion by the late Miles Pickering, we tried glyoxal as a substitute for formaldehyde. Formaldehyde is a known carcinogen, and glyoxal is a safer substitute. We could not get a polymerization reaction to occur between phenol and glyoxal, but we were successful when glyoxal and resorcinol were used with an acidic catalyst. Similarities in the structure of formaldehyde and glyoxal, and also phenol and resorcinol, are shown:

formaldehyde   glyoxal   phenol   resorcinol

▲   The reaction between resorcinol and glyoxal is probably similar to that between phenol and formaldehyde. When acid is used to catalyze the reaction, a linear soluble polymer called a *novolac* is formed. The aromatic phenol rings are linked together by methylene, $-CH_2-$, bridges that occur randomly in the ortho and para positions.

▲   Formation of the linear polymer is called *condensation polymerization*; it results in the elimination of some molecule—in this case, water.

▲   In the final stage of polymer formation, a three-dimensional cross-linked polymer is formed. This polymer is called a *thermoset resin* because it hardens irreversibly—it cannot be softened again—and it is not soluble.

▲   In the linear polymer stage, about 8% of the phenolic resins are produced as alcoholic solutions. These are basically varnishes and are called *laminating resins*. They are used to produce decorative laminates for countertops, such as Formica, and industrial laminates for electrical parts. Any volatile compounds are removed by drying in an air oven before the final processing.

▲ Glyoxal was first synthesized by Debus in 1856 but was not used until almost 80 years later. Around 1937 Peacock Laboratories used glyoxal as a reducing agent for silver in a spray process in which silver was applied to glass and plastic surfaces.

Reactions

For the reaction between formaldehyde and phenol, first a novolac is formed and then, in the final stage, a thermoset resin. The structures are shown below:

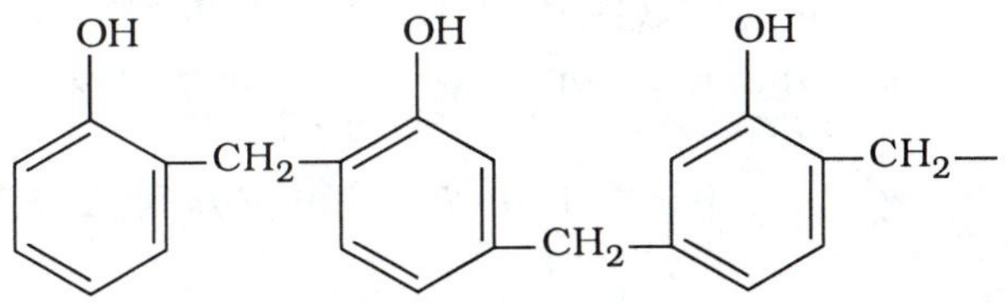

linear polymer (novolac)

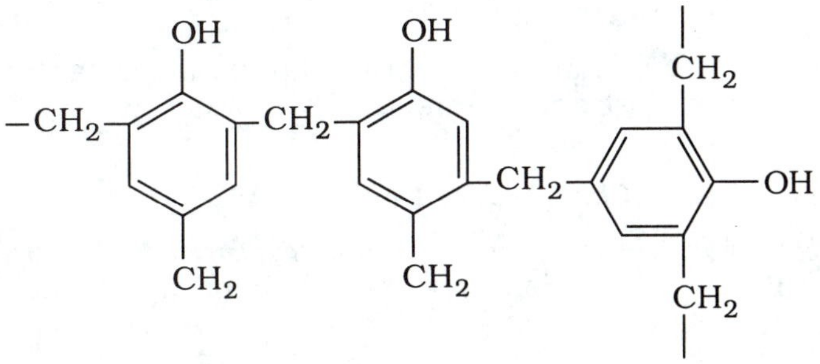

three-dimensional cross-linked polymer (thermoset resin)

Notes

1. Never use a solution of glyoxal that is stronger than commercial glyoxal, 40%. Glyoxal can be absorbed through the skin and is also flammable. A precipitate may form in the glyoxal

solution because of prolonged storage, but it can be redissolved by warming to 50–60 °C.

2. Using the specified amounts, the entire reaction could be done in a test tube of about 40-mL capacity, but the test tube will then have to be broken to remove the polymer.

3. The reaction might spurt out of the beaker, so make sure no one is standing nearby. It is also very exothermic, that is, it produces a lot of heat.

4. If difficulty is encountered with the polymer hardening, set the disposable plastic beaker in a hot water bath.

References

Billmeyer, F. W., Jr. *Textbook of Polymer Science*, 3rd ed.; Wiley-Interscience: New York, 1984; pp 437–440.

Debus, H. *Justus Liebigs Ann. Chem.* **1856**, *100*, 5.

Marvel, C. S. *Introduction to Organic Chemistry of High Polymers;* Wiley, New York, 1959; pp 24–25.

McNamee, R. W.; Barry, R. P. *Ind. Eng. Chem.* **1951**, *43*, 786–794.

Peacock, W. U. S. Patent 2,363,354, Nov. 21, 1944.

Ravve, A. *Organic Chemistry of Macromolecules;* Dekker: New York, 1967; Chapter 17.

Rosato, D. V.; Rosato, D. V. *Plastics Processing Data Handbook;* Van Nostrand-Reinhold: New York, 1990; pp 22–23.

Sears, J. A. *J. Chem. Educ.* **1971**, *48*, A499.

Tested Demonstrations in Chemistry, 6th ed.; Alyea, H. N.; Dutton, F. B., Eds.; Journal of Chemical Education: Easton, PA, 1965; p 110.

Appendices

APPENDIX 1

Topic Matrix

Investigation Number	Physical Properties	Reactions of Elements	Reactions of Gases	Energy Changes	Solutions & Solubility	Transition Elements	Kinetics & Equilibrium	Acids & Bases	Oxidation & Reduction	Electro-chemistry	Synthesis	Organic Chemistry	Historical Notes
1	•												•
2	•	•										•	
3	•	•						•				•	
4	•												
5		•			•				•				
6		•										•	•
7		•											
8		•			•		•		•				
9		•			•				•				
10	•	•	•			•			•		•		
11		•				•			•		•		
12	•	•		•			•		•		•	•	
13	•	•				•			•	•			
14	•	•			•	•	•	•	•	•			
15		•			•				•		•		
16		•			•				•	•			•

Investigation Number	Physical Properties of Elements	Reactions of Elements	Reactions of Gases	Energy Changes	Solutions & Solubility	Transition Elements	Kinetics & Equilibrium	Acids & Bases	Oxidation & Reduction	Electro-chemistry	Synthesis	Organic Chemistry	Historical Notes
17	•			•									
18			•				•		•		•		
19			•				•					•	
20			•					•				•	
21			•										
22				•							•		
23				•									•
24				•									
25				•									
26				•			•		•			•	
27					•	•							
28					•								•
29					•								
30					•								
31					•								•
32					•								•
33					•								•
34					•					•			
35					•					•			
36					•								
37					•								•
38	•				•								
39	•				•								
40						•			•				

Investigation Number	Physical Properties of Elements	Reactions of Elements	Reactions of Gases	Energy Changes	Solutions & Solubility	Transition Elements	Kinetics & Equilibrium	Acids & Bases	Oxidation & Reduction	Electro-chemistry	Synthesis	Organic Chemistry	Historical Notes
41					•	•							•
42	•				•	•							
43					•	•	•						
44	•		•		•	•			•				•
45	•					•			•	•	•		
46					•	•			•		•	•	
47	•	•		•		•			•		•		
48					•	•	•		•				
49					•	•						•	
50							•					•	
51							•					•	•
52							•					•	•
53							•		•				•
54							•	•					
55			•			•	•	•	•			•	
56							•	•			•	•	
57		•					•		•				
58								•		•			
59								•				•	
60								•				•	
61								•				•	
62						•		•					
63													
64	•	•		•	•	•	•	•	•	•	•		

Investigation Number	Physical Properties	Reactions of Elements	Reactions of Gases	Energy Changes	Solutions & Solubility	Transition Elements	Kinetics & Equilibrium	Acids & Bases	Oxidation & Reduction	Electro-chemistry	Synthesis	Organic Chemistry	Historical Notes
65							•	•				•	
66									•				•
67									•		•	•	•
68							•		•			•	•
69									•				•
70									•				
71									•			•	•
72					•	•			•				
73						•			•				
74						•			•				
75	•	•					•	•	•			•	
76										•			
77										•			
78										•			
79	•	•			•		•	•	•	•			
80									•	•			
81									•		•		•
82											•		
83											•		
84						•					•	•	
85											•	•	
86				•	•	•	•		•		•		
87			•		•	•			•		•		•
88	•	•					•		•		•	•	

Investigation Number	Physical Properties of Elements	Reactions of Elements	Reactions of Gases	Energy Changes	Solutions & Solubility	Transition Elements	Kinetics & Equilibrium	Acids & Bases	Oxidation & Reduction	Electro-chemistry	Synthesis	Organic Chemistry	Historical Notes
89					●		●		●		●		●
90		●					●		●		●	●	●
91												●	●
92												●	●
93	●											●	●
94								●				●	●
95												●	
96										●		●	
97			●		●		●	●	●			●	●
98		●	●		●		●	●	●		●	●	
99		●					●	●	●		●	●	
100		●					●					●	
101				●								●	●

Chemical Index by Investigation Number

Safety and Disposal

The following information is meant to be a general guide for you to consider before performing a demonstration and disposing of the chemical compounds formed or used in each investigation.

Always wear safety goggles when you are performing a demonstration, and insist that any students assisting you wear them also. If the demonstration requires a hood or adequate ventilation, it should not be performed without it. Take special care when making solutions and diluting acids. Remember, always add acid to water when making dilutions. Do not exceed the amounts of chemicals recommended for a demonstration, and perform each demonstration only under the conditions specified in the procedure.

Check with your district science supervisor or state department of education to see which substances are on the "banned" list for your schools and whether specific directions for disposal of chemical wastes are available. The specifications and requirements vary from state to state. Generally, you can dispose of waste acids and bases as follows:

1. Neutralize acids with aqueous ammonia or a sodium bicarbonate solution and flush down the drain with copious amounts of water.
2. Neutralize bases with a citric acid solution or dilute hydrochloric acid and flush down the drain with copious amounts of water.

The following procedures for chemicals used in these demonstrations are suggested in *Prudent Practices for Disposal of Chemicals from Laboratories* (National Academy Press: Washington, DC, 1983).

Inorganic Chemicals

Chemical	*Disposal Procedure*
Barium	Precipitate as the sulfate or carbonate and dispose of in an EPA-regulated landfill.
Cadmium	Add sodium solution to the metal ion solution to precipitate the heavy metal sulfide. Make this solution basic by adding aqueous ammonia. Wash and dry the precipitated metal sulfide. Dispose of the precipitate in an EPA-regulated landfill. Flush the remaining solution down the drain.
Other cations	In small quantities, flush down the drain with excess water.
Chromate (CrO_4^{2-}, $Cr_2O_7^{2-}$)	Check for regulations in effect in your state. Usually, you can dispose of chromates by adjusting the pH to 3 with sulfuric acid, adding 50% excess sodium bisulfite until a temperature increase results, and precipitating with NaOH and disposing of the solid in an EPA-regulated landfill.
Cobalt	Same as for cadmium.
Hexacyanoferrate(III) $[Fe(CN)_6^{3-}]$	Make solution basic (pH12); add twofold excess of 30% calcium hypochlorite. Dispose several hours later with 20-fold excess of water.
Hexacyanoferrate(II) $[Fe(CN)_6^{4-}]$	Make solution basic (pH12); add twofold excess of 30% calcium hypochlorite. Dispose several hours later with 20-fold excess of water.
Lead	Same as for cadmium.
Manganese(II)	Same as for cadmium.
Mercury	Same as for cadmium.
Nickel	Same as for cadmium.
Silver	Same as for cadmium.

Organic Chemicals

The following organic compounds must be incinerated in an EPA-regulated incinerator and should **not** be flushed down the drain:

Alcohols and phenols
Aldehydes
Amides
Carboxylic acids

Esters
Halogenated hydrocarbons
Hydrocarbons
Ketones

APPENDIX 4

Properties and Preparation of Laboratory Acids and Bases

Parameter	Ammonium Hydroxide (NH_4OH)	Acetic Acid ($HC_2H_3O_2$)	Hydrochloric Acid (HCl)	Nitric Acid (HNO_3)	Sulfuric Acid (H_2SO_4)
Dilute this volume (in milliliters) of concentrated reagent to 1 L to make a 1.0 M solution.	67.5	57.5	83.0	64.0	56.0
Dilute this volume (in milliliters) of concentrated reagent to 1 L to make a 3.0 M solution	200	172	249	183	168
Dilute this volume (in milliliters) of concentrated reagent to 1 L to make a 6 M solution	405	345	496	382	336
Normality of concentrated reagent	14.8	17.4	12.1	15.7	36.0
Molecular weight	35.05	60.05	36.46	63.02	98.08
Specific gravity	0.90	1.05	1.19	1.42	1.84
Approximate percentage in concentrated reagent	57.6	99.5	37.0	69.5	96.0

Note: To make *normal* solutions, use the same amount of reagent shown except for sulfuric acid, for which you should use half the amount indicated. Example: dilute 28.0 mL of concentrated sulfuric acid to make 1 L of 1.0 N sulfuric acid solution.

Safety note: Concentrated acids and bases are corrosive. When you work with 6 M concentrations of acids or bases, wear gloves and safety goggles and work at a safety station equipped with a face shield or freestanding shield. When you dilute acids, always add the acid to water. Caution: Solutions will become hot. Use a fume hood when diluting or using concentrated acids or bases.

Light and Color

Throughout this book we have many investigations that rely on a change in color or the production of a colored solution from a colorless solution. We have attempted to describe the specifics of the color changes most appropriate to each investigation in the Concepts sections. In this appendix we will give a general overview of color and the interaction of visible light with solutions and color-producing compounds.

The color of a solid substance is a result of selective absorption. Our eyes detect photons at frequencies of light that are not absorbed by the objects we are looking at and are therefore reflected to our eyes. When visible light is absorbed by objects, most of it is transformed into vibrational energy, and the remainder can be emitted at nonvisible frequencies. Concurrently, if incident light falling on a solid contains only those frequencies absorbed by the solid, then the object will appear black, because no visible light will be reflected to our eyes. If, on the other hand, the incident light contains only those frequencies that are reflected, then none will be absorbed. A white object absorbs very little visible light. This also means that when visible light is absorbed, energy (in the form of photons) is absorbed. Thus, at the most fundamental level, something about the substance is responsible for the absorption of specific photons of light. Each specific frequency of light corresponds to a specific amount of energy, a quantum or a photon.

Very strong absorptions result from the transition of electrons in molecular orbitals from the highest occupied molecular orbital (HOMO) to the lowest unoccupied molecular orbital (LUMO). Fundamentally, absorption causes vibrational energies to increase and electrons to move from a lower-energy *ground state* to a higher-energy *excited state*. The return pathway taken to reach either a lower-frequency vibrational state or lower-energy ground state results in a specific type of emission. Energy that is absorbed is usually dissipated as vibrational energy. Larger atoms and weaker bonds absorb in the infrared region, and smaller atoms and stronger bonds move the absorption from the infrared to visible range. Thus when we describe the color of indicators we are usually talking about smaller atoms such as carbon, nitrogen, oxygen, or sulfur atoms and double bonds, specifically alternating double and single bonds—a conjugated system.

One of the more interesting noncyclic systems is that of β-carotene, as described in Investigation 100. Another way to increase the frequency at which absorption takes place is to increase the length of the conjugated system. With β-carotene the absorption is moved to higher frequencies by extending the length of the conjugated system.

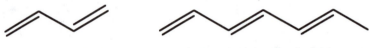

short conjugated chain   longer conjugated chain

When a nonconjugated system is changed by a chemical reaction into a conjugated system, photons are absorbed, and in the visible range, the solution will transmit colored light. Plant pigments (flavones and anthocyanins) are examples of color changes due to a structure change associated with an —O⁻ in a basic solution changing to an —OH in acid solution.

APPENDIX 6

Table of Elements

Name	Symbol	Atomic Number	Atomic Weight	Name	Symbol	Atomic Number	Atomic Weight	Name	Symbol	Atomic Number	Atomic Weight
Actinium	Ac	89	227.03	Helium	He	2	4.00	Radium	Ra	88	226.02
Aluminum	Al	13	26.98	Holmium	Ho	67	164.93	Radon	Rn	86	(222)
Americium	Am	95	(243)	Hydrogen	H	1	1.01	Rhenium	Re	75	186.21
Antimony	Sb	51	121.75	Indium	In	49	114.82	Rhodium	Rh	45	102.90
Argon	Ar	18	39.95	Iodine	I	53	126.90	Rubidium	Rb	37	85.47
Arsenic	As	33	74.92	Iridium	Ir	77	192.22	Ruthenium	Ru	44	101.07
Astatine	At	85	(210)	Iron	Fe	26	55.85	Samarium	Sm	62	150.36
Barium	Ba	56	137.33	Krypton	Kr	36	83.80	Scandium	Sc	21	44.96
Berkelium	Bk	97	(247)	Lanthanum	La	57	138.90	Selenium	Se	34	78.96
Beryllium	Be	4	9.01	Lawrencium	Lr	103	(260)	Silicon	Si	14	28.08
Bismuth	Bi	83	208.98	Lead	Pb	82	207.2	Silver	Ag	47	107.87
Boron	B	5	10.81	Lithium	Li	3	6.94	Sodium	Na	11	22.99
Bromine	Br	35	79.90	Lutetium	Lu	71	174.97	Strontium	Sr	38	87.62
Cadmium	Cd	48	112.41	Magnesium	Mg	12	24.30	Sulfur	S	16	32.06
Calcium	Ca	20	40.08	Manganese	Mn	25	54.94	Tantalum	Ta	73	180.95
Californium	Cf	98	(251)	Mendelevium	Md	101	(258)	Technetium	Tc	43	(98)
Carbon	C	6	12.01	Mercury	Hg	80	200.59	Tellurium	Te	52	127.60
Cerium	Ce	58	140.12	Molybdenum	Mo	42	95.94	Terbium	Tb	65	158.92
Cesium	Cs	55	132.90	Neodymium	Nd	60	144.24	Thallium	Tl	81	204.38
Chlorine	Cl	17	35.45	Neon	Ne	10	20.18	Thorium	Th	90	232.04
Chromium	Cr	24	51.996	Neptunium	Np	93	237.05	Thulium	Tm	69	168.93
Cobalt	Co	27	58.93	Nickel	Ni	28	58.69	Tin	Sn	50	118.69
Copper	Cu	29	63.55	Niobium	Nb	41	92.91	Titanium	Ti	22	47.88
Curium	Cm	96	(247)	Nitrogen	N	7	14.01	Tungsten	W	74	183.85
Dysprosium	Dy	66	162.50	Nobelium	No	102	(259)	Unnilennium	Une	109	(266)
Einsteinium	Es	99	(252)	Osmium	Os	76	190.2	Unnilhexium	Unh	106	(266)
Erbium	Er	68	167.26	Oxygen	O	8	15.999	Unniloctium	Uno	108	(265)
Europium	Eu	63	151.96	Palladium	Pd	46	106.42	Unnilpentium	Unp	105	(262)
Fermium	Fm	100	(257)	Phosphorus	P	15	30.97	Unnilquadium	Unq	104	(261)
Fluorine	F	9	18.998	Platinum	Pt	78	195.08	Unnilseptium	Uns	107	(262)
Francium	Fr	87	(223)	Plutonium	Pu	94	(244)	Uranium	U	92	238.03
Gadolinium	Gd	64	157.25	Polonium	Po	84	(209)	Vanadium	V	23	50.94
Gallium	Ga	31	69.72	Potassium	K	19	39.1	Xenon	Xe	54	131.29
Germanium	Ge	32	72.59	Praseodymium	Pr	59	140.91	Ytterbium	Yb	70	173.04
Gold	Au	79	196.97	Promethium	Pm	61	(145)	Yttrium	Y	39	88.91
Hafnium	Hf	72	178.49	Protactinium	Pa	91	231.04	Zinc	Zn	30	65.38
								Zirconium	Zr	40	91.22

Note: Based on carbon-12. Numbers in parentheses are mass numbers of the most stable isotopes of radioactive elements.

Subject Index

Subject Index

Copy editing: Deborah H. Steiner and Keith Ivey
Indexing: A. L. McClellan
Production: Catherine Buzzell
Acquisition: Cheryl Shanks
Cover design: Eileen Hoff

Printed and bound by United Book Press, Baltimore, MD

Notes

Notes

Bestsellers from ACS Books